Reviews and critical articles covering the entire field of normal anatomy (cytology, histology, cyto- and histochemistry, electron microscopy, macroscopy, experimental morphology and embryology and comparative anatomy) are published in Advances in Anatomy, Embryology and Cell Biology. Papers dealing with anthropology and clinical morphology that aim to encourage cooperation between anatomy andrelated disciplines will also be accepted. Papers are normally commissioned. Original papers and communications may be submitted and will be considered for publication provided they meet the requirements of a review article and thus fit into the scope of "Advances". English language is preferred.
It is a fundamental condition that submitted manuscripts have not been and will not simultaneously be submitted or published elsewhere. With the acceptance of a manuscript for publication, the publisher acquires full and exclusive copyright for all languages and countries.

Manuscripts should be addressed to
Co-ordinating Editor

215
Advances in Anatomy, Embryology and Cell Biology

More information about this series at
http://www.springer.com/series/102

Heiko Braak • Kelly Del Tredici

Neuroanatomy and Pathology of Sporadic Alzheimer's Disease

With 48 figures

Springer

Heiko Braak
Kelly Del Tredici
Zentrum f. Biomed. Forschung AG
Klinische Neuroanatomie/Abteilung Neurologie
Universität Ulm
Ulm
Germany

ISSN 0301-5556 ISSN 2192-7065 (electronic)
ISBN 978-3-319-12678-4 ISBN 978-3-319-12679-1 (eBook)
DOI 10.1007/978-3-319-12679-1
Springer Cham Heidelberg New York Dordrecht London

Library of Congress Control Number: 2014957640

Printed on acid-free paper

Springer is part of Springer Science+Business Media (www.springer.com)

Preface

The downside of the current tendency to prolonged life expectancy in developed countries is the increase in diseases associated with advanced age, especially those involving the central nervous system (CNS). Foremost among these is sporadic Alzheimer's disease (AD) which leads to dementia (Brookmeyer et al. 2007; Qiu et al. 2009; Reitz et al. 2011; Mayeux and Stern 2012). Nevertheless, today, despite all efforts on numerous fronts, no causal or disease-modifying therapy is available (Doody et al. 2014; Salloway et al. 2014). AD is a neurological disorder of the human CNS. The pathological lesions associated with the AD process require an unusually long period of time to evolve, but, in the final analysis, they result in clinically recognizable impairment of higher brain functions.

This book is written for a readership that is to some extent familiar with the anatomy of the human nervous system and is interested in the changes it undergoes during the AD process. As in the previously published book on sporadic Parkinson's disease from the same Springer series (Braak and Del Tredici 2009), the present effort approaches and interprets the pathological process in AD chiefly from a neuroanatomical perspective. However, we want to make the text readable for non-experts, *inter alia* by including throughout it both introductory and more detailed explanations pertaining to important anatomical relationships that facilitate understanding the material but that are not available in standard textbooks or only cursorily explained therein, e.g., the anatomy of the entorhinal region.

Clinically, AD only occurs in humans, and the hallmark lesions underlying the disease process predominantly are found in the human CNS. Thus, there are no truly adequate animal models for AD (Rapoport and Nelson 2011), although the implications of this reality are largely overlooked in much current research. For the past 25 years, an amyloidocentric understanding of AD research has largely ignored opposing data and arguments, thereby leaving aside important questions that still require answers (Maarouf et al. 2010). The authors focus on fundamental aspects of the AD process as a whole with the intention of encouraging alternatives to the Aβ-centered understanding of AD.

As indicated by its title, this book deals mainly with morphologically recognizable deviations from the normal anatomical condition of the human CNS. The

AD-associated pathology is illustrated from its beginnings (sometimes even in childhood) until its final form that is reached late in life. The AD process commences much earlier than the clinically recognizable phase of the disorder and its timeline includes an unusually extended non-symptomatic phase. The further the pendulum swings away from the symptomatic final stages towards the early pathology, the more obvious the lesions become, although from a standpoint of severity they are more unremarkable and, thus, frequently overlooked during routine neuropathological assessment. For this reason, we decided to deal with the hallmark lesions in early phases of the AD process in considerable detail. Clinically manifest cases of AD, on the other hand, display extensive disease-associated lesions that, as a rule, are accompanied by non-AD-related pathologies, including vascular changes and concomitant neurodegenerative disorders.

For a constitutive introduction to the morphology of AD, one of the authors (HB) owes a special debt of gratitude to an American colleague, Thomas L. Kemper, MD (Department of Anatomy and Neurobiology, Boston University School of Medicine), who also conveyed to him the fascination with the idea that AD is a disorder that adheres to the conditions of human neuroanatomy. The authors thank the Goethe University Frankfurt (The Braak Collection). They are also thankful for valuable comments provided by Khalid Iqbal, PhD (New York State Institute for Basic Research in Developmental Disabilities) and Michel Goedert, MD (MRC Laboratory of Molecular Biology, University of Cambridge). They wish to express their appreciation to Horst-Werner Korf, MD (Dr. Senckenbergische Anatomie, Goethe University, Frankfurt) for the invitation to prepare this book, Albert C. Ludolph, MD (Department of Neurology, University of Ulm) for support and helpful discussions, and Ms. Anne Clauss from Springer (Heidelberg) for careful editing of the text. They also are grateful to Jürgen Bohl, MD (formerly Department of Neuropathology, University of Mainz) for ongoing support, Ms. Simone Feldengut (Tables, silver staining, immunocytochemistry), Ms. Siegrid Baumann, Ms. Gabriele Ehmke, Ms. Julia Straub (immunocytochemistry), Mr. Hans-Jürgen Steudt (Olympus Germany, Stuttgart) for technical assistance, and Mr. David Ewert (Department of Neurology, University of Ulm) for the many hours spent preparing and helping to design the illustrations. In view of the breadth of the subject matter, it was necessary to weigh the bibliography in favor of more recent original studies and reviews. In other words, it was not the authors' intention to supply an exhaustive survey of all of the pertinent literature from the AD field.

Funding for this work was made possible, in part, by the German Research Council (Deutsche Forschungsgemeinschaft, DFG) Grant number TR 1000/1-1 and the Robert A. Pritzker Prize from the Michael J. Fox Foundation for Parkinson's Disease Research.

This book is dedicated in gratitude to the memories of Eva Braak (†2000), William R. Markesbery (†2011), and Inge Grundke-Iqbal (†2012).

Ulm, Germany
13 September 2014

Heiko Braak
Kelly Del Tredici

Contents

Chapter 1
Prologue

Sporadic Alzheimer's disease (AD) was first described by the physician Alois Alzheimer (Fig. 2.1a) as an insidious and slowly progressive neurodegenerative disorder of the human central nervous system (CNS) (Alzheimer 1906; Alzheimer et al. 1995). Clinically, its earliest sign is a subtle decline in memory functions in a state of clear consciousness. Intellectual and practical skills gradually worsen, and personality changes manifest themselves, followed by deterioration of language functions, impairment of visuospatial tasks, and, in the end, dysregulation of autonomic functions and dysfunction of the motor system in the form of a hypokinetic hypertonic syndrome (Albert et al. 2011; Morris et al. 2014). Both the tempo of the cognitive decline and the duration of the disease as well as neurological symptoms can vary considerably from one individual to another (Franssen et al. 1993).

Clinically recognizable AD develops only in humans and is found in all ethnic groups worldwide, whereby its prevalence increases remarkably with age (Purohit et al. 2011). Of known diseases causing dementia, AD is the most frequent one and accounts for 60–70 % of cases. In societies with increasing life expectancy, it can be predicted that the number of demented individuals will quadruple by the year 2050, whereby this estimate rests on the reasonable assumption that no significant therapeutic breakthrough for AD will occur in the immediate future (Brookmeyer et al. 2007; Reitz et al. 2011; Mayeux and Stern 2012). As such, AD imposes an enormous socio-political and economic threat that makes the search for causal or symptomatic solutions an urgent medical priority (Mount and Downton 2006; Trojanowski and Hampel 2011; Mayeux and Stern 2012).

At present, only a provisional clinical diagnosis can be made for AD, and this usually takes place during the last phase of the underlying pathological process. Diagnoses based on *ante mortem* observation, even with the help of advancing ancillary disciplines, such as neuroimaging and biomarkers, are still too unreliable,

H. Braak, K. Del Tredici, *Neuroanatomy and Pathology of Sporadic Alzheimer's Disease*, Advances in Anatomy, Embryology and Cell Biology 215,
DOI 10.1007/978-3-319-12679-1_1

thereby necessitating *post mortem* confirmation (Beach et al. 2012a; Montine et al. 2012; Toledo et al. 2012; Serrano-Pozo et al. 2013).

Cases of sporadic AD show the presence of the same pathological process that develops at specific CNS sites (Duyckaerts et al. 2009; Nelson et al. 2009, 2011). Currently, cross-sectional studies performed on cohorts of non-selected autopsy cases are the chief sources of information about the systematic progression of this process (Bouras et al. 1994; Giannakopoulos et al. 1994; Braak and Braak 1997b; Dickson 1997a; Duyckaerts and Hauw 1997; Braak et al. 2011). The present article is based on the results of such analyses performed on brains conventionally fixed by immersion in aqueous solutions of formaldehyde. The findings reported here involve established histological methods (immunohistochemistry, silver stains, dye-staining techniques) and can be reproduced by other laboratories.

Chapter 2
Introduction

2.1 Sporadic AD Is a Proteinopathy Linked to the Development of Intraneuronal Inclusions of Abnormal Tau Protein Which, in Later Phases, Are Accompanied by the Formation of Extracellular Plaque-Like Deposits of Amyloid-β Protein

Aggregates of abnormal proteins are the hallmarks of the AD-associated pathological process. Intraneuronal inclusions consisting of aggregated protein tau develop only in vulnerable types of CNS nerve cells (Fig. 2.1b) (Goedert and Spillantini 2006; Goedert et al. 2006; Mandelkow et al. 2007; Alonso et al. 2008; Iqbal and Grundke-Iqbal 2008; Iqbal et al. 2009; Kolarova et al. 2012; Mandelkow and Mandelkow 2012; Spillantini and Goedert 2013) whereas plaque-like deposits containing amyloid-β protein (Aβ) appear in the extracellular space of the CNS (Fig. 2.1c) (Beyreuther and Masters 1991; Selkoe 1994, 2000; Mattson 2004: Masters and Beyreuther 2006; Haass and Selkoe 2007; Alafuzoff et al. 2009; Haass et al. 2012; Masters and Selkoe 2012; Selkoe et al. 2012). Both types of proteinaceous deposits appear at different times within different CNS predilection sites and, from there, spread out systematically into previously uninvolved brain regions. In the course of the pathological process, both types of lesions increase in severity. The AD process begins with intraneuronal aggregations of the protein tau followed, after a time-lag of approximately a decade, by the extracellular deposition of Aβ (Braak and Braak 1997b; Duyckaerts and Hauw 1997; Silverman et al. 1997; Braak et al. 2013) (Fig. 9.16, Tables 9.2 and 9.6). Neuritic plaques (NPs), i.e., combined deposits consisting of insoluble Aβ with dystrophic neurites that contain aggregated tau, develop only in the late phase of the disease process (Braak and Braak 1997b; Braak et al. 2013).

H. Braak, K. Del Tredici, *Neuroanatomy and Pathology of Sporadic Alzheimer's Disease*, Advances in Anatomy, Embryology and Cell Biology 215,
DOI 10.1007/978-3-319-12679-1_2

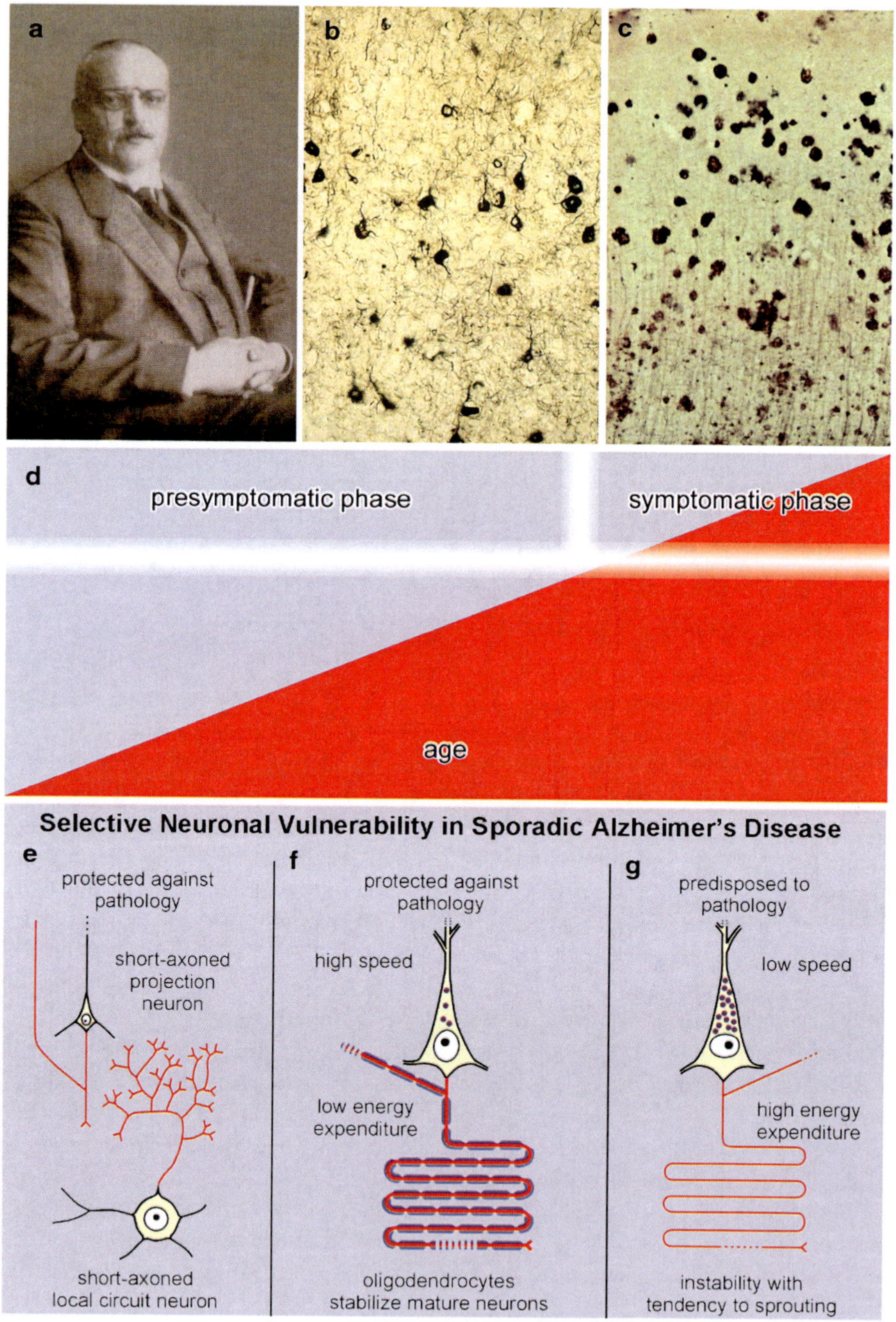

Fig. 2.1 (**a**) Alois Alzheimer. (**b**) Neurofibrillary lesions shown by Gallyas silver-iodide staining technique. (**c**) Aβ plaques visualized by Campbell-Switzer silver-pyridine staining technique. (**d**) The lesions (*red*) increase in severity and extent without remission during a long presymptomatic disease phase that reaches a shorter and final symptomatic phase. (**e–g**) Selective neuronal vulnerability in sporadic AD. Neuronal types protected against AD-associated pathology include

In contradistinction to the pathology that emerges in the course of sporadic Parkinson's disease (PD) (Braak and Del Tredici 2009; Del Tredici et al. 2010; Del Tredici and Braak 2012), the hallmark lesions associated with AD remain almost exclusively confined to the CNS. Nerve cells of the peripheral and enteric nervous systems (PNS, ENS) likewise contain normal tau and the amyloid precursor protein (APP), but only the olfactory epithelium is known to develop abnormal protein aggregates (Arnold et al. 2010; Kovács 2013). Why is it that the ENS and other PNS sites do not develop aggregated tau and Aβ? Attempts to understand the pathogenesis of the hallmark lesions should also explain why the AD process develops almost exclusively within the CNS.

The AD-associated pathological process, once started, is not known to regress, improve spontaneously, or go into remission. This controversial but important aspect of the AD process is treated below in Sect. 2.3. The process spans nearly a lifetime unless it is interrupted by death from other causes. In other words, it encompasses a much longer time-span than the clinically recognizable phase and consists of a relatively short symptomatic final phase and a long preclinical phase. The last phase, accompanied by the loss of cognitive and executive functions, manifests itself, as a rule, only at an advanced age (Fig. 2.1d).

The hallmark lesions are generally accompanied in this late phase by pathologies attributable to disorders that develop with increasing age (e.g., vascular disease, metabolic syndrome, concomitant neurodegenerative diseases) that aggravate, to varying degrees, the clinical picture, making the diagnosis and the questions surrounding the pathogenesis of AD increasingly complex (van Gool and Eikelenboom 2000; Iadecola 2004, 2010; Fotuhi et al. 2009; Duyckaerts et al. 2009; Grinberg and Thal 2010; Chui et al. 2012; Hunter et al. 2012; Korczyn et al. 2012; Korczyn 2013; Thal et al. 2012; Wang et al. 2012; Kovács et al. 2013; Serrano-Pozo et al. 2013). The co-occurrence of sporadic AD and the lesions associated with sporadic PD or frontotemporal lobe degeneration (FTLD) is especially problematic (Duyckaerts et al. 2009; Mesulam et al. 2014). Nevertheless, owing to the co-occurrence of multiple pathologies, each case is unique. For this reason, sporadic AD is not always viewed as a single disease entity but as a syndrome leading to dementia (Korczyn 2013).

In youth and in early adulthood Aβ plaque deposition is non-existent or rare. Tau aggregates, on the other hand, occur before puberty and are absent only in very young children (Fig. 9.12) (Braak et al. 2011; Braak and Del Tredici 2011, 2012). The ongoing development of both the intraneuronal tau aggregates and the extracellular plaque-like Aβ deposits is extraordinarily slow, so that the hallmark lesions cannot be said to originate only in old age or typically during aging (Morrison and Hof 1997; Nelson et al. 2011). Nonetheless, clinically overt AD has been viewed as

Fig. 2.1 (continued) short-axoned projection cells and local circuit neurons (**e**) and pyramidal cell with a long and heavily myelinated axon (**f**). Vulnerable types of pyramidal cells have a long and sparsely myelinated axon (**g**). (**e–g** adapted with permission from H Braak and K Del Tredici, Adv Anat Embryol Cell Biol 2009;201:1–119)

a disorder that is intrinsic to the aging process or, at the very least, indirectly attributable to it. Age-related factors capable of damaging aging postmitotic cells, such as chronic inflammation, oxidative stress, mitochondrial and metabolic dysfunction, blood-brain barrier impairment, or failure of the ubiquitin-proteasomal system, are viewed as pivotal factors in the pathogenesis of AD (de la Monte and Tong 2013; Pohanka 2013; Yan et al. 2013; Arshavsky 2014). None of these factors alone, however, suffices to explain why the hallmark lesions consistently and selectively develop in only a few types of nerve cells. Obviously, they fail to affect *all* known types of postmitotic cells inside and outside of the CNS. Even when the discussion is confined to the CNS, it is evident that AD does not involve *all* neuronal types there indiscriminately. Rather, the AD-process is a remarkably selective one in that it develops in only a minority of neuronal types while sparing all of the rest. In addition, the effects of age-related factors do not explain why the tau lesions develop in children and young adults during the early phase of the disease. Thus, advanced age *per se* is not necessary for the formation of the AD-related tau lesions: The pathological process underlying sporadic AD is not 'age-dependent' but an uncommonly protracted and progressive process that frequently extends into old age (Fig. 2.1d). Clinical symptoms develop subtly, and nerve cell impairment leads to a gradual loss of fundamental functions that first appear after a given threshold is exceeded (Fig. 2.1d).

The present monograph rests upon the assumption that it is the pathological process depicted above which, in its final phase, causes the clinical symptoms of AD. Individuals with a history of cognitive dysfunction, whose brains do not have the hallmark AD lesions, should be classified in the heterogeneous group of non-AD dementias (Tolnay and Probst 1999; Clavaguera et al. 2013a; Dickson et al. 2007).

2.2 Some Neuronal Types Exhibit a Particular Inclination to the Pathological Process While Others Show a Considerable Resistance To It

The deterioration of the CNS in sporadic AD specifically targets predilection sites in select subcortical nuclei and cortical areas. Of the numerous neuronal types within the CNS, only some develop AD-related pathology, whereas others, including those directly in the vicinity of involved nerve cells, remain morphologically and functionally intact (Braak and Braak 1999). The resultant non-random regional distribution pattern of the lesions is reflected by dysfunction of select neuronal types and, in some regions, mild neuronal loss. The pathological process selectively affects high-order processing regions (Arendt 2000). However, because these regions are not absolutely essential for survival the disease process can exist and progress for nearly an entire lifetime until death, which is usually caused by severe autonomic dysregulation.

Most of the vulnerable neuronal types are phylogenetically late-appearing elements that also achieve functional maturity late in life (Rapoport 1988, 1989; Bartzokis 2004; Rapoport and Nelson 2011). These nerve cells frequently retain a high degree of structural plasticity in the adult brain and show signs of immaturity that endure well into adulthood (Stephan 1983; Rapoport 1990, 1999; Arendt et al. 1998; Arendt 2000; Bufill et al. 2013). The distalmost segments of their dendrites often display a slow and ongoing growth pattern that persists even when dendritic maturation in non-susceptible nerve cell types is already complete (Arendt 2000). In addition, the vulnerable cells frequently display a loss of regulation over neuronal differentiation with a partial reactivation of the cell cycle (Arendt 2004, 2005, 2012).

Nearly all of the diverse types of nerve cells in the CNS that are prone to develop AD-related tau aggregates are projection neurons, i.e., nerve cells with a localized dendritic arbor and an axon that is disproportionately long in relation to the size of the host cell soma (Fig. 2.1f, g). Inasmuch as glutamatergic, gabaergic, dopaminergic, noradrenergic, serotonergic, histaminergic, and cholinergic projection cells become involved, the type of neurotransmitter or neuromodulator synthesized is not essential for predicting which nerve cells are especially vulnerable or predisposed to the AD process. Short-axoned neurons generally do not become affected (Fig. 2.1e) (Braak and Braak 1999; Benarroch 2013) with the rare exception of large cholinergic local circuit neurons in the striatum and chandelier cells in the basal temporal neocortex (see Sect. 9.2). In addition, nearly all projection neurons with a short axon remain intact, such as the small pyramidal cells in neocortical layers II and IV (spiny stellate cells) (Fig. 2.1e) (DeFelipe et al. 2002) and those of the presubicular parvocellular layer. An aggrecan-based perineuronal net surrounds subsets of cortical and subcortical projection neurons, and these subsets do not develop intraneuronal tau lesions. As a result, perineuronal nets may contribute to the selective resistance of these nerve cells against the tau-mediated pathological process (Morawski et al. 2010). The overall resistance on the part of short-axoned neurons has its consequences: When a projection essentially synapses only on local circuit neurons—e.g., as is the case with neocortical pyramidal projections from layer V to layer III and those from layer VI to layer IV (Bannister 2005)—the pathological process cannot spread any further. The potential route of propagation from projections of the anterior subnuclei of thalamus to the presubicular parvocellular layer also comes to a halt at the short-axoned neurons there.

All of the endangered neuronal types generate a long and thin-caliber axon that either is encased by a thin myelin sheath or does not undergo myelination (Fig. 2.1g). Projection neurons reach full functional maturity only after their axons have achieved their stipulated degree of myelination (van der Knaap et al. 1991). Human neocortical projection neurons in prefrontal or in high-order sensory association areas commence myelination late in life and, thus, are thinly myelinated and predisposed to the AD-related process (Bartzokis 2004). By contrast, cortical or subcortical projection neurons with heavily myelinated axons resist developing tau aggregates (Fig. 2.1f). These include the Betz cells in the primary motor area, Meynert's pyramidal cells in the striate area, or host neurons of the medial longitudinal fascicle.

Increased thickness of the myelin sheath provides greater velocity of axonal conduction and, at the same time, a considerable reduction of the metabolic demands placed on the host cell for the transmission of the impulses (Fig. 2.1f). By contrast, rapid-firing projection neurons with an unmyelinated or immaturely myelinated axon are subject to higher energy turnovers and are thereby chronically exposed to oxidative stress (Fig. 2.1g) (Pohanka 2013; Yan et al. 2013). The relatively postponed onset of myelogenesis in neurons of high-order regions which are not absolutely essential for preservation of basic brain functions results in high energy consumption (Fig. 2.1g). Yet, it is precisely these neurons that enrich and optimize complex activities, such as learning, memory, and perception, that are particularly prone to develop AD lesions (Arendt 2000). These include late-maturing pyramidal cells, whose axons preferentially develop connections to the distalmost dendritic segments of other cortical pyramidal cells but have no immediately obvious functions.

The myelin sheath provides a mechanical barrier against viruses and other pathogens by virtually isolating the axon from the surrounding extraneuronal space. However, this also means that the energy supply for long axons cannot originate in the cell soma. Instead, glial cells increasingly assume this function. The axon is embedded among oligodendroglial cells that protect and support it. Local astrocytes take up substances critical for the energy balance via contacts to the cerebral vasculature and redistribute them to oligodendrocytes by means of gap junctions. Additional mechanisms enable the transfer of these substances from the oligodendrocyte to the axon (Nave 2010; Lee et al. 2012b).

Nerve cells of the human adult generally are richly supplied with lipofuscin or neuromelanin granules (Braak 1980; Double et al. 2008) and, notably, all of the vulnerable cell types contain such granules (Fig. 2.1g). The presence of lipofuscin or neuromelanin deposits alone, however, does not suffice to account for the susceptibility of projection cells to the AD-process because many of them develop large amounts of these paraplasmatic granules with advancing age *without* developing tau aggregates, as, for instance, the Betz cells of the motor cortex or projection neurons of the lateral geniculate body. On the other hand, nerve cells that conspicuously lack lipofuscin or neuromelanin granules, or that contain only a few granules even in old age, consistently resist the pathological process. Prime examples are the large projection neurons of the hypothalamic lateral mamillary nucleus where, even in old age, lipofuscin granules are hardly present and AD-associated tau aggregates do not develop.

Viewed against this background, the combination of paraplasmatic pigment granules and an unmyelinated or sparsely myelinated long and thin-caliber axon in phylo- and ontogenetically late-maturing projection neurons appears to be a deficiency of the human CNS that may be necessary for the induction of the AD-associated process (Fig. 2.1g) (Braak and Braak 1999).

2.3 Consistent Changes in the Regional Distribution Pattern of Intraneuronal Inclusions Make a Staging Procedure Possible

As in other illnesses, at some point during the pathological process, patients cross a threshold from the preclinical phase to the symptomatic manifestation of AD (Fig. 2.1d). By the time clinicians make their diagnosis, patients are, relatively speaking, in the late phase of a larger pathological process. The disease festers in the CNS for decades until its dimensions are such that dysfunctional behavior becomes manifest (Dubois et al. 2010).

Cases with clinically recognizable symptoms usually can be assigned to one of four neuropathological subgroups (neurofibrillary stages III, IV, V, VI), which differ with respect to the topographic extent of the AD-related tau pathology (Fig. 2.2a, e). The idea of regional expansion rests on the assumption that the lesions most likely develop in a consecutive manner within the CNS and then increase in severity and extent (Fig. 2.2b). Each subgroup, therefore, displays newly affected regions in addition to the tau lesions existing at previously involved sites (Braak et al. 2011).

Tau aggregates occur as incidental findings in non-symptomatic individuals (Linn et al. 1995; Dubois et al. 2010; Ferrer 2012), and these also can be divided into one of four subgroups (Fig. 2.2c, d). These lesions sometimes are viewed as a variant of neuronal aging, neuroprotective, or as possible markers of a non-AD-related tauopathy (Attems et al. 2012; Cower and Mudher 2013; Jack and Holtzman 2013; Jack et al. 2013; Thal et al. 2013; Braak and Del Tredici 2014; Crary et al. 2014; Kuchibhotla et al. 2014). However, they also can be interpreted as markers of early phase disease—comparable to malignant cells in a carcinoma that fail to produce symptoms but mark the onset of a pathological process (Fig. 2.2e). The concept that incidental tau aggregates are completely 'normal' requires the highly problematic definition of the point at which such tau inclusions convert from a 'normal' status into 'disease-related' lesions. Disease-related lesions existing prior to the clinical manifestation of a disease are usually regarded as prodromes. As a clinical entity, AD includes dementia, but the AD-related pathological process includes a very protracted preclinical phase, which certainly occupies a pivotal position in relation to the pathogenesis of AD (Figs. 2.1d and 2.2e).

We view such clinically mute incidental tau lesions as a potential threat to the CNS for the following reasons: The presymptomatic and symptomatic disease phases are both marked by the presence of the same types of intraneuronal tau aggregates in the same types of nerve cells and at the same regional predilection sites. Second, since the lesional pattern of the last preclinical subgroup closely resembles that of the first symptomatic subgroup (compare Fig. 2.2d and b), both sets of subgroups combined can be taken to reflect the full spectrum of the pathological AD-process (Fig. 2.3a–c). The lesions develop in a remarkably predictable and consistent sequence across cases (Figs. 2.2e, 9.8, and 9.13) (Kemper 1984; Arnold et al. 1991; Braak and Braak 1991a, 1997a, b, 1999; Braak

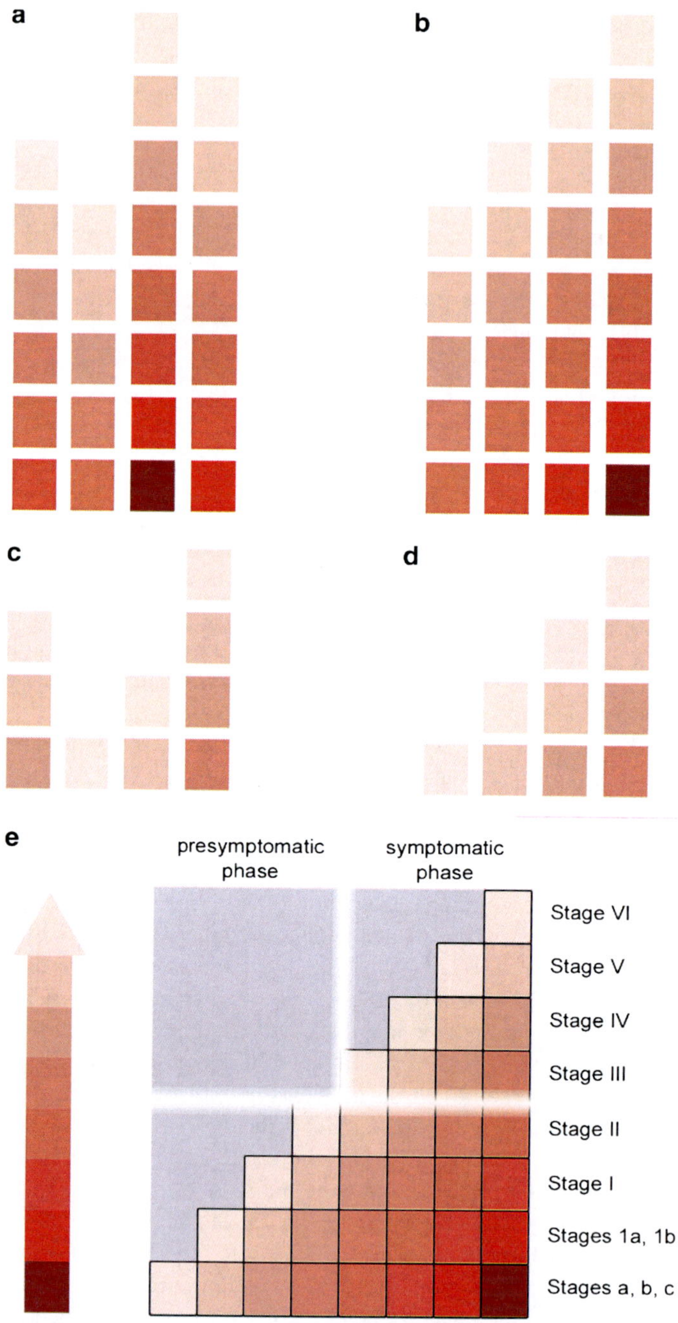

Fig. 2.2 Presymptomatic and symptomatic phases of sporadic AD. (**a**) Most symptomatic cases with AD-associated tau pathology fall into four subgroups. (**b**) Given the consistency of this finding and, based on the topographic distribution pattern of the lesions, the four groups can be

et al. 2006a; Hyman and Gómez-Isla 1994; Duyckaerts and Hauw 1997; Delacourte et al. 1999). Third, the existence of asymmetric tau distribution patterns, as seen in double hemispheres sections immunostained for AT8, indicates that a neurobiological continuum of tau pathology exists and that this pathology also tends to progress within one and the same individual, albeit at a different pace (Fig. 2.4a–c).

Locating the first tau aggregates in an organ as voluminous as the human CNS might seem like an insurmountable undertaking, but it is possible provided the predilection sites of the pathological process are known. The lesions do not develop randomly, here at one site, there at another. Instead, the AD process is stereotypic, beginning in the same regions and advancing with little inter-individual variation. As a rule, one does not see appreciable differences bilaterally with respect to the topographic distribution of tau pathology in double hemisphere sections (Fig. 2.3), and when discrepancies occur, they do not amount to more than one (Figs. 2.3a and 2.4a) or two stages (Fig. 2.4b, c). In cases with advanced stages, this phenomenon disappears and the hemispheres of such cases display a more or less symmetric involvement (Fig. 2.3c). In this context, it should be noted that the concept of neuropathological staging is recommended for practical reasons only. In principal, it is an artificial construct because, as pointed out above, the hallmark lesions develop continually rather than in definite steps (Braak and Braak 1991a).

The AD-associated process begins with the appearance of non-argyrophilic tau lesions in stages a-c in brainstem nuclei that diffusely project to the cerebral cortex and progresses from there into the cerebral cortex, i.e., the transentorhinal region (non-argyrophilic cortical lesions in stages 1a and 1b) (Fig. 5.3). Thereafter, the inclusions partially convert into argyrophilic lesions that characterize NFT stages I–VI. From the transentorhinal region (stage I), the pathology advances into the entorhinal region and hippocampal formation (stage II) (Fig. 7.3). During stage III, tau lesions encroach upon the adjoining basal temporal neocortex (Fig. 9.2). In stage IV, they extend more widely into the temporal neocortex, insula, subgenual and anterogenual frontal regions, and anterior cingulate areas (Fig. 9.8). Stage V cases display severe involvement of most neocortical association areas, leaving only first-order association areas and primary fields mildly involved or intact. Finally, nearly all cortical areas show neurofibrillary changes in stage VI (Table 2.1; Figs. 2.2e, 2.3c, and 9.13) (Braak and Braak 1991a; Braak et al. 2006a). Abnormal tau formation continues to take place from the beginning (stage a) until the end

Fig. 2.2 (continued) arranged to show disease progression (neurofibrillary stages III, IV, V, VI). (**c**) Similarly, most non-symptomatic cases also fall into four subgroups (**d**) and can be ordered sequentially (pretangle stages a–c, 1a/1b, and NFT stages I, II). (**e**) Mild to moderate tau lesions develop over time until a threshold from the prodromal to the symptomatic (clinical) phase is crossed. Roman numerals correspond to stages of Gallyas-positive (argyrophilic) lesions. Arabic numerals represent stages with cortical AT8-ir (non-argyrophilic) lesions (stages 1a/1b), whereas lower case letters designate stages with subcortical AT8-ir non-argyrophilic lesions (stages a–c). See also Table 2.1

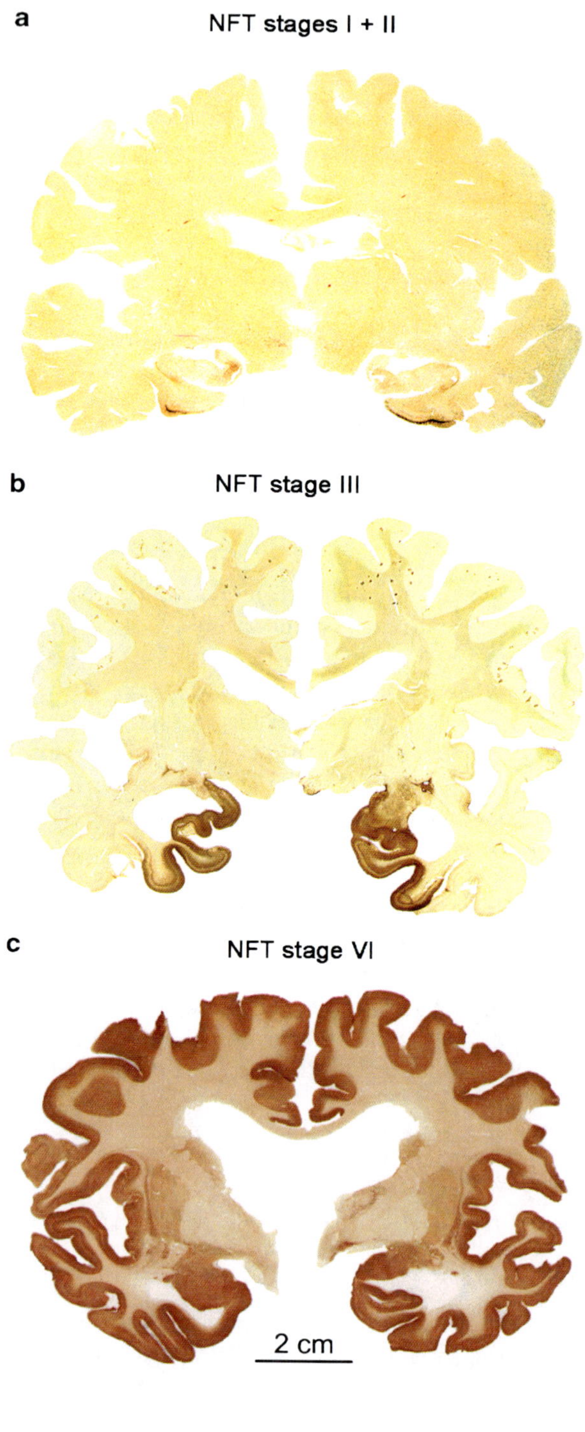

Fig. 2.3 Overview of AD-related AT8-immunopositive tau stages in three double hemispheres of 100 μm thickness. (**a**) The hemispheres of this cognitively intact 80-year-old female display NFT stages I (*left*) and II (*right*) in the absence of Aβ plaques. (**b**) NFT stage III in a 90-year-old female, who died of a malignant pancreatic neoplasm. Aβ plaques were also present. (**c**) Frontal section from a severely demented 72-year-old female AD patient (cause of death aspiration pneumonia) with bilateral NFT pathology and ventricular widening typical of stage VI. Aβ plaques were also present (adapted, in part, with permission from H Braak and K Del Tredici, Alzheimers Dement. 2012; 8:227–233). Scale bar in (**c**) is valid for (**a**) and (**b**)

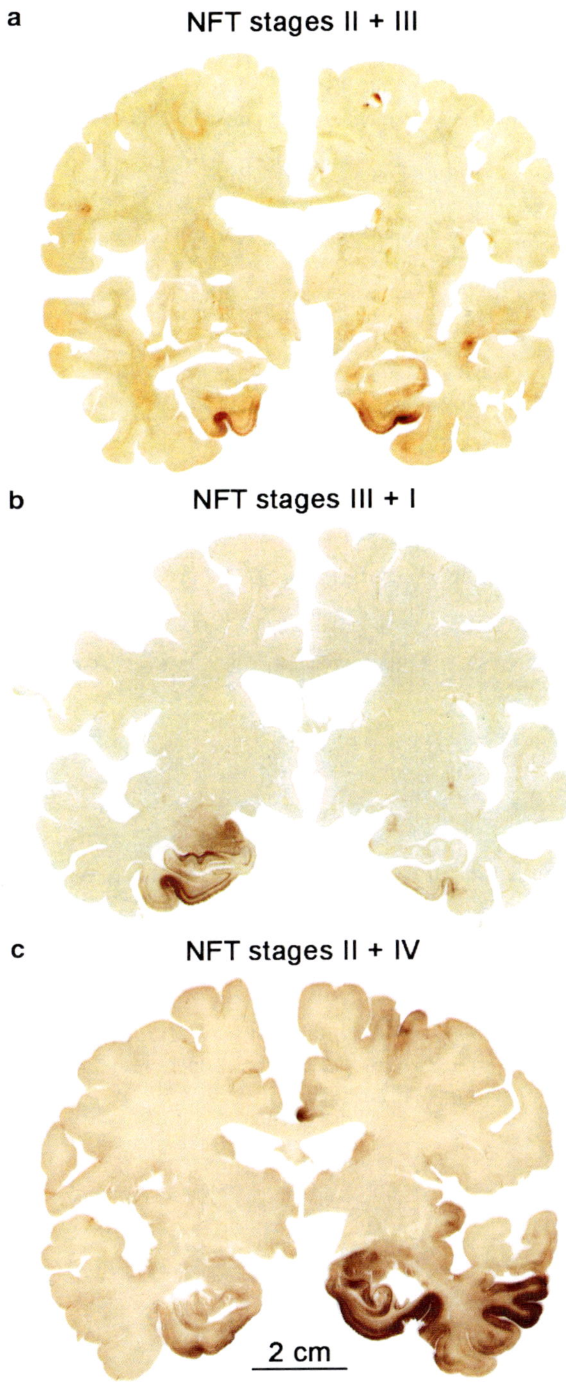

Fig. 2.4 Double hemispheres of 100 μm thickness showing asymmetrical AD-related AT8-immunopositive tau stages. (**a**) Section from a cognitively intact 77-year-old female (cause of death myocardial infarction) at NFT stage III (*left*) and stage II (*right*). No ventricular widening is detectable. (**b**) Frontal section from a 75-year-old female patient (cause of death chronic lymphatic leukemia) with NFT stages III (*left*) and NFT stage I (*right*). (**c**) This frontal section from a cognitively impaired 75-year-old male (cause of death metastatic pulmonary neoplasm) displays pathology corresponding to NFT stages II (*left*) and IV (*right*). Deviations of more than one NFT stage are unusual (compare Fig. 2.3). See also Sect. 9.3. Scale bar in (**c**) applies also to (**a**) and (**b**)

Table 2.1 Overview of CNS sites where intraneuronal inclusions sequentially develop during sporadic AD

Stage a	AT8-immunopositive projection neurons (proximal axon only) of the locus coeruleus
Stage b	AT8-immunopositive projection neurons (axon and somatodendritic compartment) of the locus coeruleus
Stage c	AT8-immunopositive neurons of other non-thalamic nuclei with diffuse cortical projections
Stage 1a	AT8-immunopositive nerve cell processes in the transentorhinal region (cerebral cortex)
Stage 1b	AT8-immunopositive modified pyramidal cells (layer pre-α cells) of the transentorhinal region
Stage I	Gallyas-positive (argyrophilic) pyramidal cells, predominantly in the transentorhinal region, olfactory bulb, and anterior olfactory nucleus; chiefly AT8-immunopositive axons and α-motoneurons in the spinal cord
Stage II	Gallyas-positive pyramidal cells, predominantly in the entorhinal region
Stage III	Gallyas-positive pyramidal cells in layers pre-α and pri-α of transentorhinal/entorhinal regions, pyramidal cells in either the first or second Ammon's horn sector, pyramidal cells in basal temporal neocortex
Stage IV	Gallyas-positive pyramidal cells in the first and second Ammon's horn sectors, mild involvement of third and fourth sector, temporal and frontal neocortex (including middle temporal gyrus, insula, and subgenual, anterogenual, anterior cingulate areas of the frontal lobe)
Stage V	Gallyas-positive granule cells of the dentate fascia and pyramidal cells of the subiculum. Gallyas-positive pyramidal cells in virtually all prefrontal areas and high order sensory association areas
Stage VI	Gallyas-positive pyramidal cells in all first order sensory association areas, premotor areas, primary sensory areas, finally in the primary motor area. Severe involvement of dentate granule cells and subicular pyramidal cells

stage of the pathological process. There is no evidence that the pathological process halts its development or 'burns out.'

In other words, the pathological process selectively involves a few neuronal types in the CNS and it progresses from distinct predilection sites in the brainstem to the transentorhinal cortex. Little by little, in a sequential and orderly pattern, it encompasses previously uninvolved regions, above all in the cerebral cortex. It is primarily the pronounced directional expansion of the pathological process that requires an explanation (Table 2.1).

Chapter 3
Basic Organization of Non-thalamic Nuclei with Diffuse Cortical Projections

The pattern of lesions that slowly develops during the AD-process is more easily understood and interpreted against the anatomical backdrop of involved regions and their major connections, which can be depicted in simplified diagrams (Figs. 3.1a, 6.9, and 6.10).

A few nuclei within the brainstem, midbrain, basal forebrain, and hypothalamus send long and profusely ramifying projections to the olfactory bulb, the cerebral cortex, many subcortical nuclei (with the notable exception of the pallidum), the cerebellum, and spinal cord. Because of their extensive projections, these nuclei belong to a functionally unified group, the non-thalamic nuclei with diffuse cortical projections, in contrast to the nuclei of the thalamus that project to specific cortical areas (Fig. 6.10).

Alone the fact that the axons of projection neurons from the non-thalamic nuclei have such an unusually large number of ramifications (Fig. 5.3a) indicates that they are not destined to relay data selectively from one site to another, but that their effect on cortical regions is more universal and generalized (Morrison et al. 1982). To this end, such axons share a common hallmark, namely, their terminal branches develop numerous local varicosities equipped only with presynaptic sites (non-junctional varicosities) supplemented by varying amounts of classical synaptic connections that have both presynaptic and postsynaptic sites (Agnati et al. 1995; Nieuwenhuys 1999; O'Donnell et al. 2012). By means of the non-junctional varicosities, the axons release neurotransmitters and neuromodulators diffusely into the interstitial fluid (ISF) (volume transmission), thereby activating receptors of astrocytes, oligodendrocytes, and microglial cells, as well as of cells of the vasculature and nerve cells within a given local diffusion zone (Kalaria et al. 1989; Agnati et al. 1995; O'Donnell et al. 2012). The mechanism of volume transmission permits influence of neurotransmitters and neuromodulators on many cells within a given local diffusion zone.

H. Braak, K. Del Tredici, *Neuroanatomy and Pathology of Sporadic Alzheimer's Disease*, Advances in Anatomy, Embryology and Cell Biology 215,
DOI 10.1007/978-3-319-12679-1_3

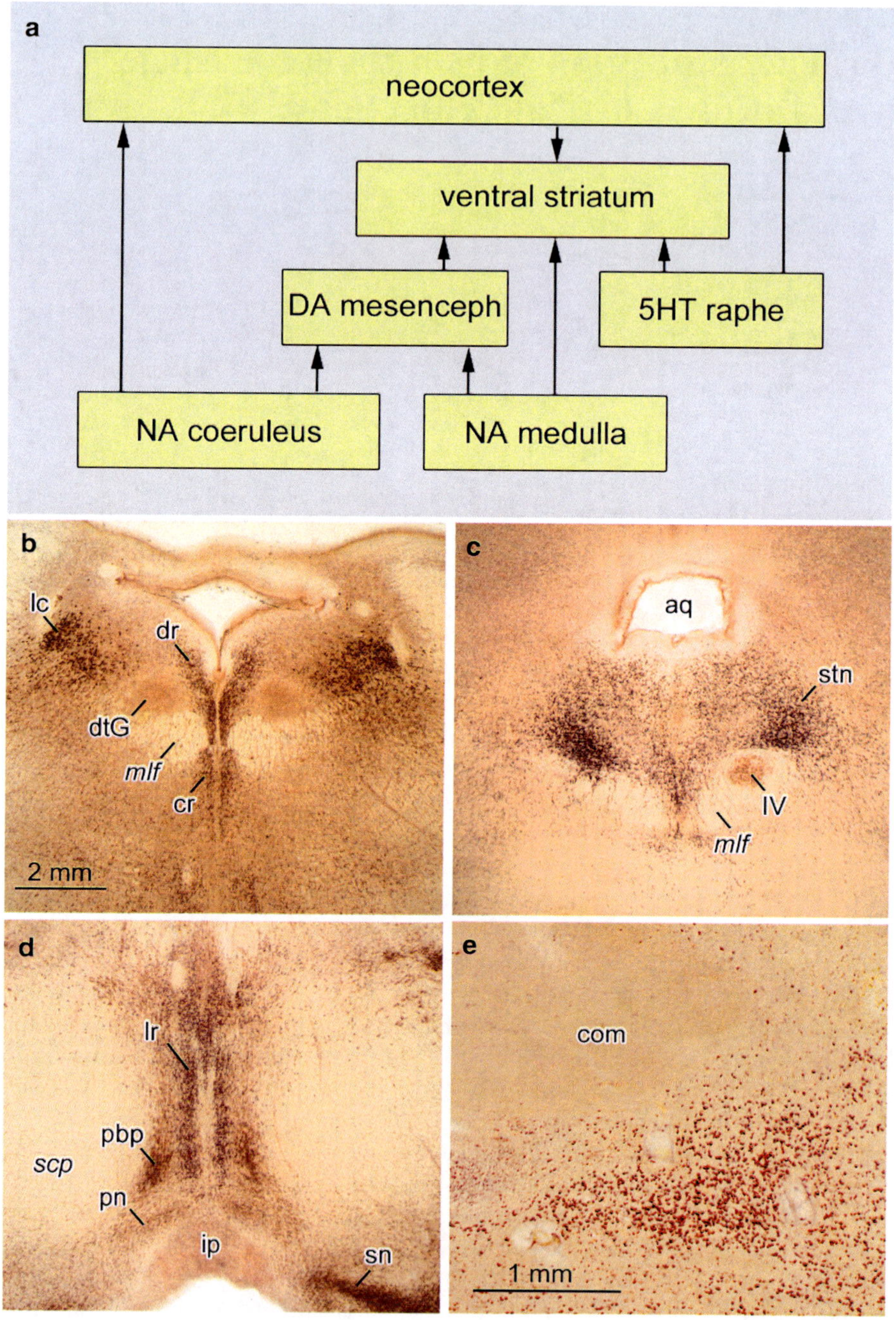

Fig. 3.1 (**a**) Diagram of interconnections between important non-thalamic nuclei with diffuse projections to the cerebral cortex. (**b–e**) 400 μm sections cut perpendicularly to the brainstem axis. Pigment-Nissl stain (aldehydefuchsin and *Darrow red*) from a 53-year-old male control. (**b**) Locus coeruleus and dorsal raphe nucleus. (**c**) Supratrochlear portion of dorsal raphe nucleus and motor nucleus of IVth cranial nerve in the medial longitudinal fascicle. (**d**) Paranigral nucleus,

Non-thalamic nuclei with diffuse cortical projections include not only the noradrenergic locus coeruleus (Fig. 3.1a, b) and related areas but also the serotonergic nuclei of the upper raphe system (Fig. 3.1a–c), the dopaminergic paranigral and parabrachial pigmented nuclei of the mesencephalic tegmentum (Fig. 3.1a, d), the histaminergic tuberomamillary nucleus of the hypothalamus (Panula et al. 1990; Saper 2004), and the cholinergic magnocellular nuclei of the basal forebrain (Fig. 3.1e) (Mesulam 2004).

All of these nuclei belong to the isodendritic core of the brain (Ramón-Moliner and Nauta 1966) and are located far apart from each other. During the later phases of primate evolution when the cerebral cortex increased considerably in size, the nuclei under consideration developed new subnuclei as extensions of their phylogenetically older portions. An example of this trend to 'integrated phylogeny' (Rapoport 1988, 1989) in humans is the supratrochlear portion of the dorsal raphe nucleus, which, alone by virtue of its size, is very prominent (Fig. 3.1c) and thereby differs from corresponding subnuclei in the brains of non-human mammals.

The locus coeruleus in the pontine tegmentum contains nearly 32,000 nerve cells and is the largest accumulation of noradrenergic neurons in the human brain (Aston-Jones and Cohen 2005; Szabadi 2013). Mature projection neurons in this nucleus display deposits of neuromelanin granules. The lipofuscin-like matrix of these granules becomes melanized by taking up by-products of the catecholamine metabolism (Double et al. 2008). Their natively brown color makes it possible to identify clusters of melanized neurons located just below the surface of the fourth ventricle, and they are also clearly visible in cross-sections through the brainstem (Fig. 3.1b). The slender column-like principal portion of the locus coeruleus lies close to the lateral angle of the fourth ventricle (Fig. 3.1b **lc**), commences from the level of the VIIth cranial nerve, and extends rostrally to approximately the level of the decussation of the IVth cranial nerve. The locus coeruleus receives relatively few afferents, some from nearby nuclei (for instance, from the prepositus nucleus of the hypoglossus), whereas other afferents originate in the contralateral locus coeruleus, and important steering projections come from the central subnucleus of the amygdala. Additional afferents have their origins in distant regions, e.g., the modulating projections from dorsal areas of the prefrontal neocortex to the locus coeruleus (Benarroch 2009; Sara 2009).

The subcoeruleus portion underneath the anterior half of the principal portion is less densely packed with melanized neurons and generates mainly descending

---

Fig. 3.1 (continued) parabrachial pigmented nucleus, and linear raphe nucleus. (**e**) Basal nucleus of Meynert. Abbreviations: **aq**—mesencephalic aqueduct; **com**—anterior commissure; **cr**—central raphe nucleus; **DA mesenceph**—dopaminergic neurons, mesencephalon; **dtG**—dorsal tegmental nucleus of Gudden; **dr**—dorsal raphe nucleus; **ip**—interpeduncular nucleus; **lr**—linear raphe nucleus; ***mlf***—medial longitudinal fascicle; **NA coeruleus**—noradrenergic neurons, locus coeruleus; **NA medulla**—noradrenergic neurons, medulla oblongata; **pbp**—parabrachial pigmented nucleus; **pn**—paranigral nucleus; ***scp***—decussation of superior cerebellar peduncle; **sn**—substantia nigra; **stn**—supratrochlear portion of the dorsal raphe nucleus; **5HT**—serotonergic neurons; **IV**—motor nucleus of the trochlear nerve. Scale bar in (**b**) is valid for (**c**) and (**d**)

projections. The flattened cerebellar portion projects to the cerebellum and merges with the principal portion extending along the superior cerebellar peduncle directly underneath the roof of the fourth ventricle. Closely related to the locus coeruleus are more loosely arranged groups of melanized neurons elsewhere in the lower brainstem: one group is located in lateral portions of the intermediate reticular zone. A second group is a component of the dorsal vagal area, and a third group consists of noradrenergic cells surrounding the facial nucleus and the complex of the superior olive (Counts and Mufson 2012).

The serotonergic upper raphe complex comprises the dorsal, central, and linear nuclei. The dorsal nucleus lies between the medial longitudinal fascicles and the ependymal lining of the fourth ventricle (Fig. 3.1b **dr**). It almost extends as far caudally as the locus coeruleus. Rostrally, it covers the motor nucleus of the trochlear nerve (Fig. 3.1b **IV**) and forms the expansive supratrochlear subnucleus (Fig. 3.1c **stn**). The central nucleus lies ventrally from the medial longitudinal fascicles (Fig. 3.1b **cr**), whereas the linear nucleus extends far into the decussation of the superior cerebellar peduncle (Fig. 3.1d **lr**).

With the exception of melanized neurons in the substantia nigra, the dopaminergic neurons of the midbrain also belong to this system of non-thalamic nuclei with diffuse cortical projections. The perirubral subnucleus lies close to the red nucleus, while the paranigral nucleus (Fig. 3.1d **pn**) forms an arch covering the interpeduncular nucleus (Fig. 3.1d **ip**) and continues into the leaf-like and sagittally-oriented pigmented parabrachial nucleus (Fig. 3.1d **pbp**) that accompanies the linear raphe nucleus (Fig. 3.1d **lr**).

The histaminergic/gabaergic neurons of the tuberomamillary nucleus extend through posterior tuberal and anterior mamillary territories of the phylogenetically young lateral hypothalamus and are closely associated with the median forebrain bundle. The tuberomamillary nucleus surrounds the globular cellular islands of the lateral tuberal nucleus almost in their entirety. This complex of the lateral hypothalamus is remarkably expansive in the human brain (Braak and Braak 1992b).

Clusters of large and chiefly cholinergic multipolar neurons located in the basal forebrain form three major nuclei: the medial septal nucleus, the interstitial nucleus of the diagonal band, and the basal nucleus of Meynert (Figs. 3.1e and 6.2 **bnM**) (Sassin et al. 2000; Mesulam 2004; Zaborsky et al. 2008; Schliebs and Arendt 2011; Geula and Mesulam 2012). Leaf-like extensions of the basal nucleus impinge deeply on the external and internal medullary layers of the pallidum and form the peripallidal subnucleus. The cholinergic forebrain nuclei are viewed as a relay system between components of the autonomic system, the limbic circuit, and the neocortex (Heimer and van Hoesen 2006).

Mature projection neurons of all of these nuclei have a medium-sized to large rounded soma with a marginally placed nucleus. Strongly staining Nissl material is concentrated at peripheral portions of the soma and in invaginations of the nuclear envelope. The cells contain large deposits of either lipofuscin or neuromelanin granules (Fig. 3.1). The rounded pigment deposits extend with offshoots into the proximal dendrites but not into the axon hillock or axon. Many of the lipofuscin-laden cell types contain small acidophilic granules within their somata (Issidorides et al. 1991). A pale central region of the soma remains nearly devoid of Nissl material and paraplasmatic granules.

The projection neurons of all these nuclei generate long, thin, and sparsely myelinated axons that ramify very extensively and terminate profusely in the cerebral cortex and in many additional sites (Figs. 3.1a and 5.3a). However, despite their highly ramified state, these axons do not project to the same degree to all regions of the forebrain. In this context, it is altogether remarkable that the components of the entire pallidum are not connected to the cholinergic, serotonergic, and noradrenergic networks. Again, taken together, nuclei with diffuse projections are not absolutely essential for preservation of basic brain functions, i.e., for sustaining life. Nevertheless, they influence generalized cortical functions: foremost arousal, attentiveness, ideation, and cognitive performance. Via the volume transmission of their neurotransmitters and neuromodulators, they regulate the CNS microcirculation and functions of the blood brain barrier (Reinhard et al. 1979; Kalaria et al. 1989). Collectively, they boost and enhance performance within all regions of the CNS where their axons terminate (Raichle et al. 1975; Usher et al. 1999; Farkas and Luiten 2001; Iadecola 2004, 2010; España and Berridge 2006; Hamel 2006; Samuels and Szabadi 2008: Benarroch 2009; O'Donnell et al. 2012; Chamberlain and Robbins 2013; Toussay et al. 2013).

Chapter 4
Microtubules and the Protein Tau

The protein tau is central to the pathological process underlying sporadic AD. It is chiefly a neuronal protein and is produced in all portions of the nervous system (CNS, PNS, ENS) (Trojanowski et al. 1989). In immature nerve cells, tau is distributed diffusely throughout the entire cell, but during cell maturation it converts into a primarily axonal protein. Tau originates in the cell soma's rough endoplasmatic reticulum, it clears all cellular checkpoints for quality control, and, in this fully functional state, is then transferred to the axonal compartment (Lee et al. 2001; Scholz and Mandelkow 2014). In healthy nerve cells, the protein promotes self-assembly of axonal microtubules and stabilizes them (Fig. 4.1a). Together with motor proteins, microtubules are required for anterograde and retrograde transport of substances within the axon. To a lesser extent, the protein is also present in neuronal somata, dendritic processes, astrocytes, and oligodendrocytes. However, in contrast to the situation in axons, the microtubules in the somatodendritic compartment are mainly stabilized by other microtubule-associated proteins (Mandelkow and Mandelkow 1998, 2012; Binder et al. 2004; von Bergen et al. 2005; Ávila 2006; Goedert et al. 2006; Alonso et al. 2008; Iqbal et al. 2009).

In the CNS of the human adult, six isoforms of tau are generated from a single gene on chromosome 17q21.31 by alternative mRNA splicing. The isoforms range from 381 to 441 amino acids in length and, in the PNS, they are supplemented by a "big tau" with 695 amino acids. The six isoforms differ by the presence or absence of up to three inserts. In the fetal brain, only the shortest isoform is present and is more extensively phosphorylated than tau in the adult brain. Fetal tau is probably necessary for microtubules during periods of intense plasticity in early brain development. In general, nerve cells of the human adult (including those that become involved during AD) show three-repeat (3R) and four-repeat (4R) isoform levels that are in equilibrium. Other tauopathies, such as progressive supranuclear palsy (PSP), corticobasal degeneration (CBD), and argyophilic grain

H. Braak, K. Del Tredici, *Neuroanatomy and Pathology of Sporadic Alzheimer's Disease*, Advances in Anatomy, Embryology and Cell Biology 215,
DOI 10.1007/978-3-319-12679-1_4

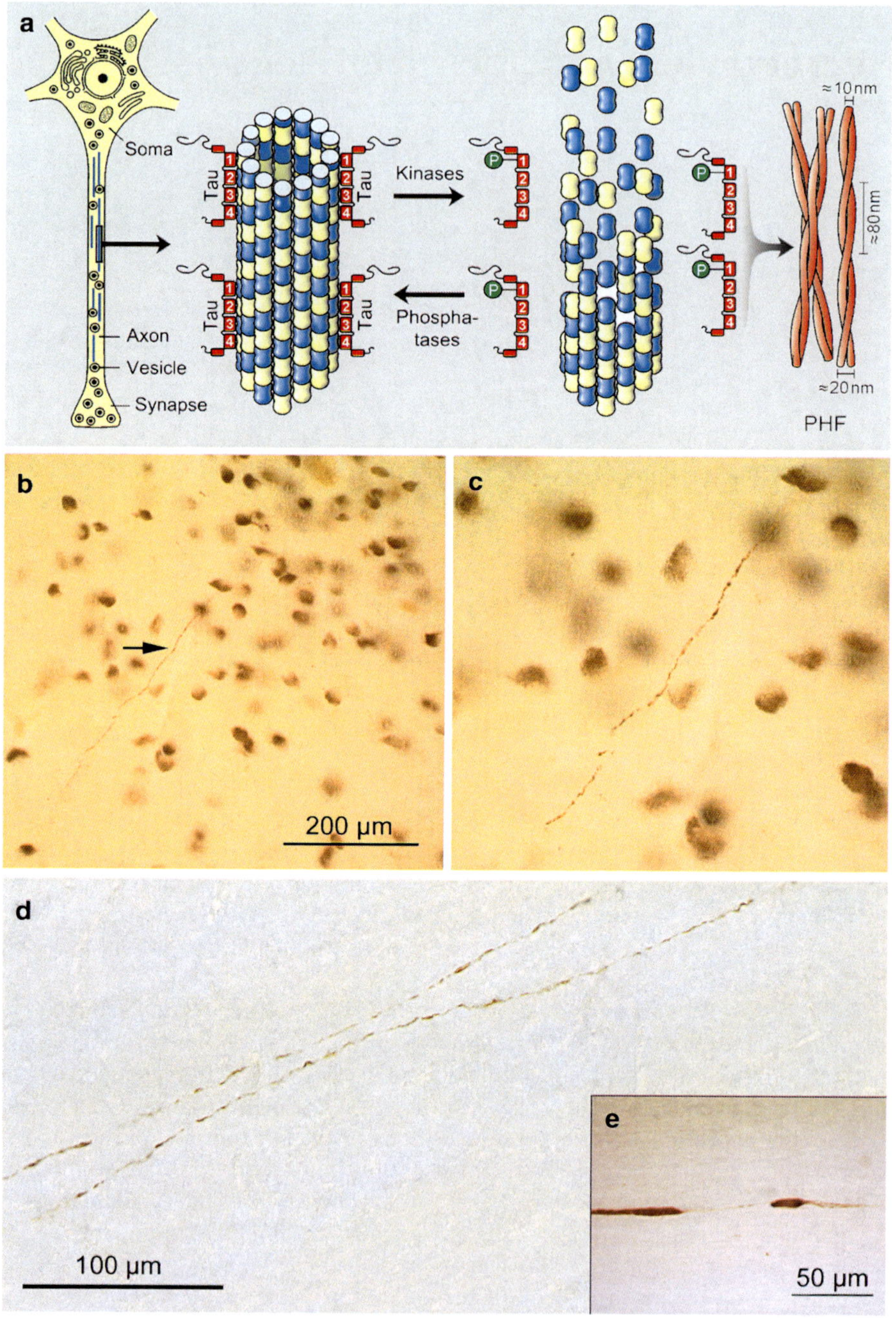

Fig. 4.1 (**a**) Schematic drawing of axonal microtuble (adapted with permission from EM Mandelkow and E Mandelkow, Trends Cell Biol 1998;8:425–427). (**b–d**) In stage a, abnormal tau occurs in axons of coeruleus projection neurons. (**b**) *Arrow* points to an AT8-ir axonal process in the locus coeruleus of a 14-year-old male shown in (**c**) at higher magnification. (**d**) Again, based

disease (AGD), exhibit a selective four-repeat accumulation, whereas tau in Pick's disease (PiD) displays a selective three-repeat isoform (Goedert et al. 1992, 1997; Buée et al. 2000; de Silva et al. 2003; Yoshida 2006; Uchihara et al. 2011, 2012; Niblock and Gallo 2012).

Consistent with its natively unfolded character, a large fraction of the tau protein consists of hydrophilic amino acids, which make it highly flexible and soluble in the cytosol. The protein has numerous binding sites at which it can undergo phosphorylation. Tau's phosphorylation status is constantly subject to fluctuation: it can transiently be up-regulated—for instance, during mitosis, anesthesia, or hibernation—without resulting in long-term harm to neuronal networks. Hyperphosphorylation of tau also occurs during early developmental phases when it enables axonal growth cones to achieve maximal flexibility for guidance of axons to their target regions and for synapse formation. Thus, an occasional and limited degree of hyperphosphorylation at appropriate sites solely on the part of monomeric tau is not an index of a disease state (Arendt 2004; Run et al. 2009; Seitz et al. 2012; Whittington et al. 2013).

Normal tau binds relatively tightly to axonal microtubules, whereas transiently hyperphosphorylated tau dissociates from microtubles and becomes redistributed freely in the axoplasma. Physiologically, a relatively stable equilibrium emerges between axonal tau microtubule binding and disengagement, thereby indicating a continuous alternation between a less highly phosphorylated state (i.e., bound to axonal microtubules) and a slightly more highly phosphorylated state (i.e., soluble in the axoplasm). This equilibrium of monomeric tau molecules is directed by local axonal kinases and phosphatases (Ballatore et al. 2007).

Monomeric hyperphosphorylated tau is highly soluble and, as such, a large proportion of the protein probably is lost during tissue fixation. It may be for this reason that currently available tau-specific antibodies preferentially mark forms of the protein that already are to some extent aggregated and slightly more viscous than monomeric tau and in this form perhaps more amenable to fixation. All of the various isoforms of tau contain about 30 or more phosphorylation sites (Goedert et al. 1992; Buée et al. 2000). Aggregated and hyperphosphorylated tau protein can be visualized with antibodies, such as the antibody AT8, which recognizes phosphate-dependent epitopes (Mercken et al. 1992). AT8-immunoreactions are highly robust, functioning reliably even in tissue obtained under suboptimal conditions, including longer post-mortal intervals or long tissue storage in aqueous solutions of formaldehyde (Pikkarainen et al. 2010). Thus, AT8 immunohistochemistry is recommended for visualizing abnormal tau during routine neuropathological assessment (Alafuzoff et al. 2006; Montine et al. 2012).

Fig. 4.1 (continued) on their length and location (here, in the superior cerebellar peduncle of a 26-year-old female), these AT8-immunoreactive (AT8-ir) processes are axons rather than dendrites. Note that within both axons non-immunoreactive segments are interspersed by AT8-ir segments (see also insert in **e**). Abbreviation: *PHF* paired helical filaments. Scale bar in micrograph (**d**) applies also to (**c**)

When it is tightly bound to microtubules, monomeric tau is protected against aggregation and, for this reason, only hyperphosphorylated tau species can aggregate. Given an initial conformational change (cross-sheet) followed by some degree of aggregation, the protein tau undergoes conversion from a normal to a pathologically altered cellular component (von Bergen et al. 2005). Here, it should again be emphasized that formation of aggregated tau is an exceptional event that only occurs in a minority of nerve cell types. It is still unclear, however, what the causes for this selectivity are, what conditions need to prevail, and which co-factors have to be present before tau aggregation begins in prediposed types of nerve cells.

One possible precondition for aggregation is a transitory peaking of high concentrations of monomeric hyperphosphorylated tau in the axoplasm. These conditions and other presently unknown factors could abruptly produce initial tau aggregates. The resultant increase in the size of the tau molecules coupled with a slight reduction in solubility, however, fundamentally alters the properties of the tau species. Such tau aggregates are not only precipitated by conventional fixation fluids and readily visualized but, more importantly, they are no longer subject to regulation by local kinases and phosphatases: that is, the abnormal material persists in its hyperphosphorylated state and becomes completely resistant to autophagy and other endogenous cellular removal mechanisms (Kovacech et al. 2010).

Thus, it is likely that it is also primarily these minimally aggregated forms of irreversibly hyperphosphorylated tau that exert a toxic or undesirable influence on their surroundings (Kopeikina et al. 2012). To evade such effects, involved neurons convert the protein as quickly as possible into larger tau aggregates that subsequently become building blocks for neurofibrillary inclusions. Formation of the AT8-immunoreactive (AT8-ir) abnormal material continues from the outset (stage a) until the end-stage (stage VI) of AD and is not known to undergo spontaneous remission (Grundke-Iqbal et al. 1986; Tolnay and Probst 1999; Trojanowski and Lee 2000; Binder et al. 2004; Goedert et al. 2006; Mandelkow et al. 2007; Kolarova et al. 2012).

Chapter 5
Early Presymptomatic Stages

5.1 Stage a: The Appearance of Abnormal Tau in Axons of Coeruleus Projection Neurons

Despite the existence of a variety of neurotransmitters and modulator substances, each of the non-thalamic nuclei with diffuse ascending projections is highly susceptible to the AD-associated pathological process (Tomlinson et al. 1981; Whitehouse et al. 1981, 1985; Mann et al. 1982; Mann 1983; German et al. 1987, 1992; Zweig et al. 1988; Chan-Palay and Asan 1989; Hertz 1989; Busch et al. 1997; Rüb et al. 2000; Sassin et al. 2000; Parvizi et al. 2001; Zarow et al. 2003; Mesulam et al. 2004; Haglund et al. 2006; Grudzien et al. 2007; Weinshenker 2008; Grinberg et al. 2009; Simic et al. 2009; Braak and Del Tredici 2011; Elobeid et al. 2011; Vana et al. 2011; Trillo et al. 2013). The long and poorly myelinated axons of these nuclei might be capable of reverting to an earlier and less well differentiated state with a higher degree of tau phosphorylation more readily than heavily myelinated axons.

The AD process begins in the locus coeruleus during childhood or in young adulthood and, from there, goes on to involve other non-thalamic nuclei with diffuse cortical projections (German et al. 1992; Benarroch 2009; Sara 2009; Braak and Del Tredici 2011). The occurrence of aggregated hyperphosphorylated tau in a single coeruleus projection neuron (stage a) (Fig. 4.1b–d) may already be sufficient to pave the way for the formation of all of the succeeding lesions (stages b–VI). Many questions that arise in this context cannot be answered yet: Why does the AD-process commence in the locus coeruleus and not elsewhere? Can the process be extinguished after a spontaneous dying off or targeted removal of the initially involved neurons? Is a continual initiation over a given period of time required to fan the disease process to such an extent that it proceeds inexorably from that point on? Does the induction of the pathological process involve mechanisms and pathways that are different from those regulating its propagation?

H. Braak, K. Del Tredici, *Neuroanatomy and Pathology of Sporadic Alzheimer's Disease*, Advances in Anatomy, Embryology and Cell Biology 215,
DOI 10.1007/978-3-319-12679-1_5

In this context, it should be pointed out that no other neuronal system develops such abundant axonal ramifications and such close contact to the microvasculature of the CNS. Coeruleus axons regulate the microvascular system and the blood brain barrier of the CNS to the same extent that the PNS does so for the macrovascular supply to the CNS—namely, via the direct release of noradrenaline from non-junctional presynaptic varicosities (Farkas and Luiten 2001). A re-uptake of the greater part of the transmitter into the presynaptic varicosities via noradrenaline transporters occurs. Pathogens, such as viruses, which reach the vasculature or the CSF, can directly influence or penetrate into axons of the brain's endogenous noradrenergic system (Pamphlett 2014).

Not all projection neurons within the locus coeruleus become involved simultaneously. Abnormal tau inclusions begin to develop within a few coeruleus projection neurons and then additional neurons follow suit. It is possible that this sequence is not fortuitous but predefined. We assume that the process begins in one or two nerve cells followed by additional cells on the opposite side. A delayed spread of the pathology to the non-involved side or an asymmetrical involvement pattern is relatively seldom (Fig. 2.4). The continual increase in the number of involved cells in the locus coeruleus occurs so gradually and slowly that one sees non-involved nerve cells even in end stage cases of AD. Contrary to what we would anticipate, given its very early involvement, the locus coeruleus does not become completely devoid of neurons during the lifelong AD-associated process.

Without light microscopically detectable precursors, slightly aggregated forms of hyperphosphorylated (AT8-ir) tau abruptly appear in the proximal axon of noradrenergic coeruleus neurons (Figs. 4.1b, c and 5.3b) (Braak and Del Tredici 2012). Their position is not irrelevant because functional limitations in the proximal axon will have a ripple effect well into the axon's terminal ramifications. Obviously, the pathological material in the proximal axon is slightly less soluble in the axoplasm than monomeric tau because it does not distribute rapidly throughout the entire axon, as might be expected were the material to consist of molecules circulating freely in the axoplasm (Maeda et al. 2007; Brunden et al. 2008). Instead, the material resembles a somewhat viscous mass that moves slowly into more distally located portions (Fig. 5.3c). During formation, neither the initial axon segment nor the axon hillock nor adjoining portions of the somatodendritic compartment display any traces of abnormal material. In other words, there is no patent evidence for a transfer of abnormal tau material from the soma of involved cells into the axon. Swellings proximal or distal to the inclusions are not seen. These findings permit the assumption that the initially altered tau material in axons originates from hyperphosphorylated tau proteins of the local axoplasm.

It is frequently reported in the literature that hyperphosphorylated tau quickly relocalizes from the axon back into the somatodendritic compartment (Thies and Mandelkow 2007; Serrano-Pozo et al. 2011). Histologically, there is no evidence for such a re-sorting phenomenon, and, in fact, it seems rather improbable that, once the pathological aggregates have so suddenly appeared in the axon, sufficient amounts of diffusible hyperphosphorylated tau would be available for such a re-location.

The altered neuronal processes are straight, thin, and thread-like, and they can be found not only in the locus coeruleus (Fig. 4.1b, c) but also in the superior cerebellar peduncle (Fig. 4.1d) and at sites where the ascending catecholaminergic tract courses through the midbrain, i.e., at sites not accessible to dendrites of coeruleus melanized neurons. Moreover, the lengths of the single AT8-ir processes exceed those of the dendrites found within the locus coeruleus (Fig. 4.1d). Taken together, these findings support the interpretation that the first cellular processes to become involved are segments of proximal axons belonging to coeruleus projection neurons.

The sudden appearance of the pathological material has consequences for the cytoskeleton of the proximal axon. Abnormal changes occur in axonal microtubules, supposedly to the point of disintegration because the partially aggregated tau no longer fulfills its primary function of stabilizing the microtubules and new monomeric tau can neither be generated quickly enough in the cell body nor be transferred into the proximal axon. As a result, anterograde or retrograde axonal transport no longer occurs effectively along such badly damaged axons (Alonso et al. 2008; Higuchi et al. 2002; Mandelkow et al. 2007; Li et al. 2007; Iqbal et al. 2009). Nonetheless, the fact remains that most involved neurons survive, by means of presently unknown mechanisms, despite the presence of severe damage. Yet, such nerve cells must also be handicapped with respect to their normal functions.

At some point, the abnormal tau inclusions in the axon convert from a viscous state into a virtually insoluble polymeric mass. The short inclusions that ultimately appear in the axon are more or less of similar length and they are interspersed by AT8-immunonegative segments (Figs. 4.1e and 5.3d). This condition is stable and persists for decades: The axon can be observed to remain intact even into the final stages of the disease process despite the existence of considerable cytoskeletal abnormalities (Velasco et al. 1998). Most probably, the insoluble inclusions consist of inert and highly aggregated forms of irreversibly hyperphosphorylated tau species since they do not change in size and do not shift in either direction, nor do they show signs of re-absorption or degradation. Within AT8-immunonegative axonal segments, immunoreactions directed against axonal cytoskeletal proteins (e.g., SMI 312) are negative. At the same time, the pronounced argyrophilia of intact axons, which can be visualized in sections stained with the Campbell-Switzer silver technique (Campbell et al. 1987; Braak and Braak 1991b) is not seen in AT8-immunonegative axonal segments.

Extensive portions of abnormally altered axons can be observed in the white substance, e.g., within bundles of the catecholaminergic tract in the medulla oblongata or throughout the course of the perforant pathway. There is no evidence that these insoluble inclusions "back up" at critical junctures within the axon, such as at branching points, and the directionality of the reduced, but probably possible, transport cannot be determined from the shape of small varicosities before or behind the inclusions. Perhaps as a reactive response, acute local destruction of the axonal cytoskeleton may secondarily and transiently induce abnormally high concentrations of newly produced tau protein within the somatodendritic domain.

An important light microscopic observation is that neither the viscous (but to some extent soluble) material nor their insoluble end-products in axons become Gallyas-positive. This general absence of argyrophilia fundamentally distinguishes abnormal material in axons from somatodendritic pretangle material that tends strongly to convert into Gallyas-positive fibrils.

It is unlikely that either the normal monomeric form of hyperphosphorylated tau or its aggregated and insoluble species are detrimental or toxic to the immediate surroundings. Noxious effects probably originate soley from the more viscous forms of tau (Brunden et al. 2008; Ding and Johnson 2008; Iqbal et al. 2009; Mufson et al. 2013; Tian et al. 2013) that display both a loss of physiological function (binding to microtubules) and a gain of toxic function (destruction of microtubules, impaired axonal transport). Host nerve cells are known to produce stable and insoluble species of the abnormal material to sequester the partially soluble and potentially dangerous species (Ittner et al. 2011). Thus, the abnormal and subsequently insoluble axonal inclusions remain *in situ* until their host neurons die while physically restricting the residual functions of axonal transport. The precise mechanisms and pathways by means of which vital transport functions in abnormally altered axons are preserved are unknown. However, residual functioning makes prolonged survival possible for many badly damaged nerve cells. The altered axon disintegrates only when the host neuron is lost, and the insoluble non-argyrophilic material then is degraded *in situ* without leaving behind any traces.

5.2 Stages b and c: Pretangle and Tangle Material Develops in the Somatodendritic Compartments of Coeruleus Neurons and Similar Lesions Appear in Additional Brainstem Nuclei with Diffuse Cortical Projections

The formation of abnormal material, initially similar to that observed in the axon, develops next in the somatodendritic compartment of coeruleus neurons, although the sources of this pretangle material cannot be convincingly identified at present. Owing to the development of the axonal inclusions, the axon lacks sufficient amounts of normal tau, and this situation possibly stimulates renewed tau production in the soma. However, because too few binding sites are available in the soma the newly synthesized tau probably lingers for a time in a hyperphosphorylated state in the cytoplasm. Under these conditions, it is conceivable that maximal concentrations can easily be exceeded, which then facilitate the formation of tau aggregates (pretangle material). The semantics of the term 'pretangle material' presume that a conversion into filamentous and argyrophilic inclusions (i.e., dendritic neuropil threads and somatic neurofibrillary tangles) is anticipated. For this reason, the term 'non-argyrophilic abnormal' tau is intentionally applied to the inclusions in axons that resemble pretangle material without undergoing such a conversion.

The first observable abnormal change in the somata of coeruleus projection cells consists of a transient appearance of droplet-like and probably viscous pretangle material (Figs. 5.2a–c and 5.3e) that appears to be partially soluble and, for this reason, may be highly toxic. The diameters of the droplet-like inclusions vary but clearly surpass those of somatic lipofuscin or neuromelanin granules. Large portions of the soma then become filled with this abnormal material, which surrounds both Nissl bodies and clusters of pigment granules without recognizably displacing these cellular components. The viscous pretangle material frequently spares the pale central portion of the soma. In dendritic processes, it first appears in distal branching points. From there, it fills the proximal dendritic portions and then merges with the viscous material in the soma (Fig. 5.3f).

Provided the pretangle tau persists and the involved cells do not die, the abnormal material in the somatodendritic portion tends to undergo conversion into strongly argyrophilic filamentous lesions. In addition, the material is tagged with ubiquitin, a heat shock protein required for the non-lysosomal ATP-dependent breakdown of abnormal proteins (Weaver et al. 2000; von Bergen et al. 2005; Ballatore et al. 2007; Maeda et al. 2007). Given the fact that the somatodendritic material is already aggregated at this point, so that it cannot be eliminated by the proteasome system (Kovacech et al. 2010), this 'delayed' ubiquitination seems futile and only ends up costing the neuron wasted energy (Garcia-Sierra et al. 2003; von Bergen et al. 2005).

Ubiquitination probably occurs shortly before the pathological material becomes argyrophilic (Lasagna-Reeves et al. 2012). The argyrophilic inclusions are selectively positive using specific silver staining techniques, such as the silver-iodide method proposed by Gallyas, which detects the lesions with great clarity (Gallyas 1971; Braak and Braak 1991b; Uchihara et al. 2001, 2005; Binder et al. 2004; Braak et al. 2006a). Neuropil threads (NTs) fill up the dendrites (Fig. 5.3h) (Ashford et al. 1998), whereas neurofibrillary tangles (NFTs) develop in the cell bodies (Fig. 5.3f, g).

There are still no answers to the question why this sequence of events is confined to the very few predisposed neuronal types described above (Table 2.1). In addition, it is unclear why involved nerve cells cannot immediately eliminate the abnormal material at a time when the molecules in question are still subject to proteasomal recycling and autophagy. Nevertheless, during this interval the viscous pretangle material remains untagged by ubiquitin (Bancher et al. 1989; Köpke et al. 1993; Uchihara et al. 2001). A dismantling or degradation of abnormal tau or signs for cellular distress, such as swelling of the neuronal soma or displacement of the Nissl material or the cell nucleus to the periphery are not observed. In other words, cell death is not imminent. On the contrary, involved nerve cells tolerate their insoluble pretangle material and, subsequently, their argyrophilic NTs/NFTs remarkably well. Thus, the viscous abnormal material is acutely toxic only initially and only briefly, whereas the insoluble material is not incompatible with long-term neuronal survival, although survival is not synonymous with optimal cell functioning.

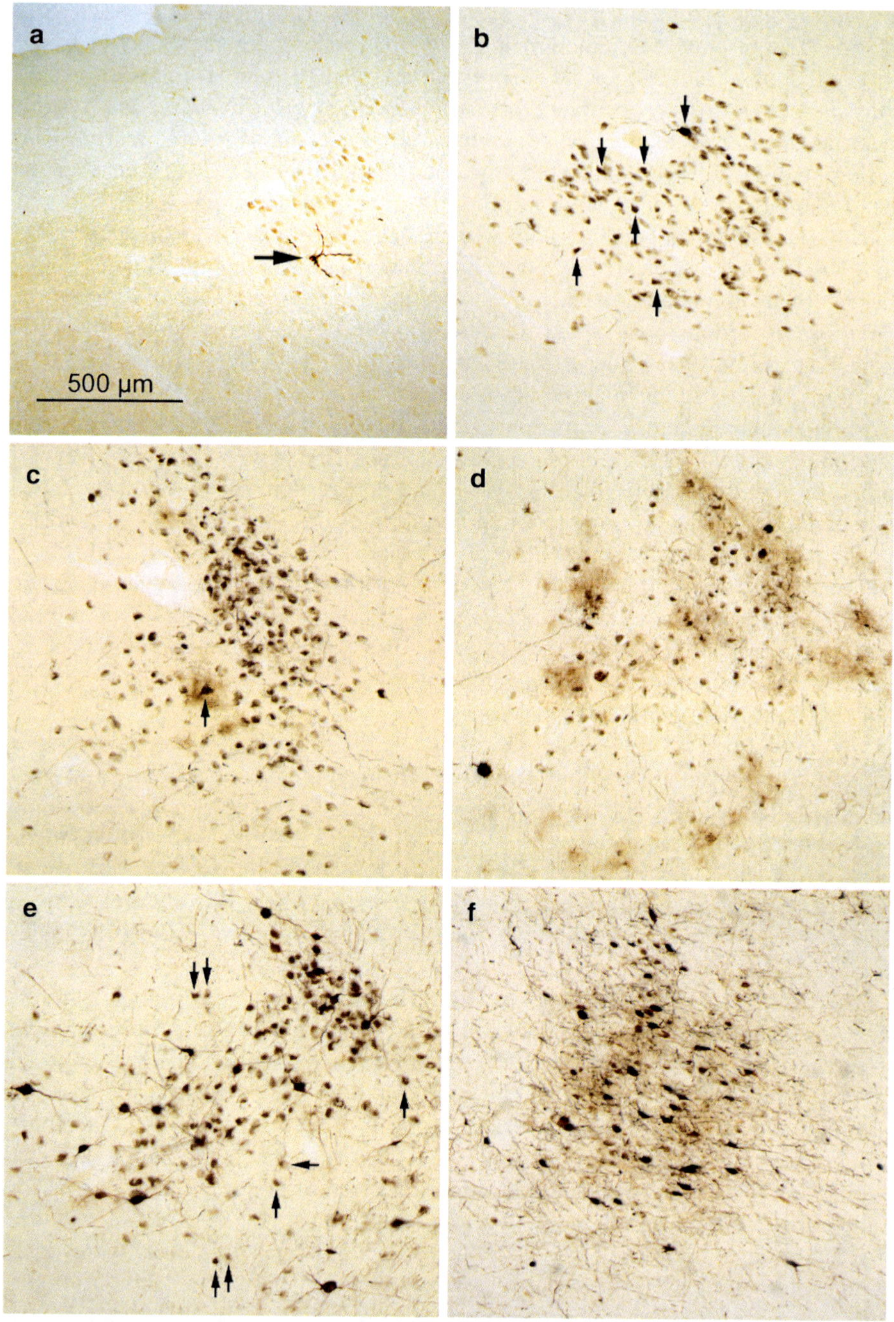

Fig. 5.1 Stages of AT8-immunopositive tau pathology in the locus coeruleus. (**a**) A single affected projection cell (*arrow*) seen in an 11-year old male (stage b). (**b**) Stage 1a is characterized by involvement (*arrows*) of multiple projection neurons (49-year-old female). (**c**) Affected projection cells and a single involved astrocyte (*arrow*) in a 68-year-old male at NFT stage I. (**d**) Multiple immunoreactive astrocytes (*brownish* 'clouds') and neuronal loss in 72-year-old

Melanized nerve cells of the locus coeruleus that survive the first phase of pretangle material formation produce additional AT8-ir material at branching points of their dendrites. Thereafter, this material protrudes into other portions of the dendrites. Eventually, bubble-like AT8-ir varicosities develop within the dendrites (Figs. 5.2g–i and 5.3g). From these varicosities, a transition becomes evident to increasingly distorted dendrites with spherically shaped and intensely AT8-ir portions interspersed with thin thread-like segments (Figs. 5.2g and 5.3h). Distal portions of such dendrites often occur as globular structures isolated in the neuropil and separated from their stems (Fig. 5.2n). A probable consequence of these abnormalities in the dendritic trees of locus coeruleus projection neurons is the severe loss of the afferences projecting to this region, foremost among which are those of the regulatory projections from the prefrontal cortex. As a rule, such severely altered neurons die and ultimately leave behind a tombstone tangle (Fig. 9.3g, h) (Augustinack et al. 2002).

Physiologically healthy melanized locus coeruleus neurons have smoothly contoured cell bodies and only a few, relatively long dendrites (Fig. 5.3e). However, from subcortical stage b onwards, the outer somatic rim of involved cells begins to develop unusual spiked protrusions (Figs. 5.2d–f and 5.3g). These newly formed spike-like structures are slender dendritic processes with gradually diminishing diameters that are tightly packed with pretangle material (Fig. 5.2e). It is possible that melanized neurons of the locus coeruleus can eliminate some of the pathological material and/or other non-degradable inclusions (e.g., neuromelanin granules) via these spiked protrusions. Neuromelanin granules within the locus coeruleus are normally confined only to nerve cells and, for this reason, it is noteworthy that in this nucleus they are, nonetheless, also encountered extraneuronally, spaced widely apart from each other. The existence of a specialized exocytotic mechanism would explain the frequent presence of extraneuronal neuromelanin granules in young individuals.

From stage III onwards, AT8-ir changes nearby abnormal dendrites in the locus coeruleus develop in the form of a corona of thin astrocytic processes fanning out like the spokes of a wheel around the altered dendrites (Figs. 5.1c, d, 5.2o–q, and 5.3h, i). The distal portions of these astrocytic processes are enlarged and have direct contact with the pathologically altered dendritic segments. They are also intensely AT8-ir whereas the astrocytic somata are immunonegative or show only a weak reaction. As such, it appears that a degradation of the abnormal pretangle material occurs within the astrocytic processes without leaving behind any visible remnants.

Similar corona-like accumulations of astrocytic processes are seen in the locus coeruleus surrounding dying or already avital nerve cells, whose somata are almost

---

Fig. 5.1 (continued) male at NFT stage II. (**e**) Intact projection cells (*arrows*) amidst severely involved neurons accompanied by increasingly heavy neuronal loss in an 86-year-old female AD patient (NFT stage V) and (**f**) in an 80-year-old female with AD (NFT stage V). 100 μm sections. Scale bar in (**a**) also applies to micrographs in (**b–f**)

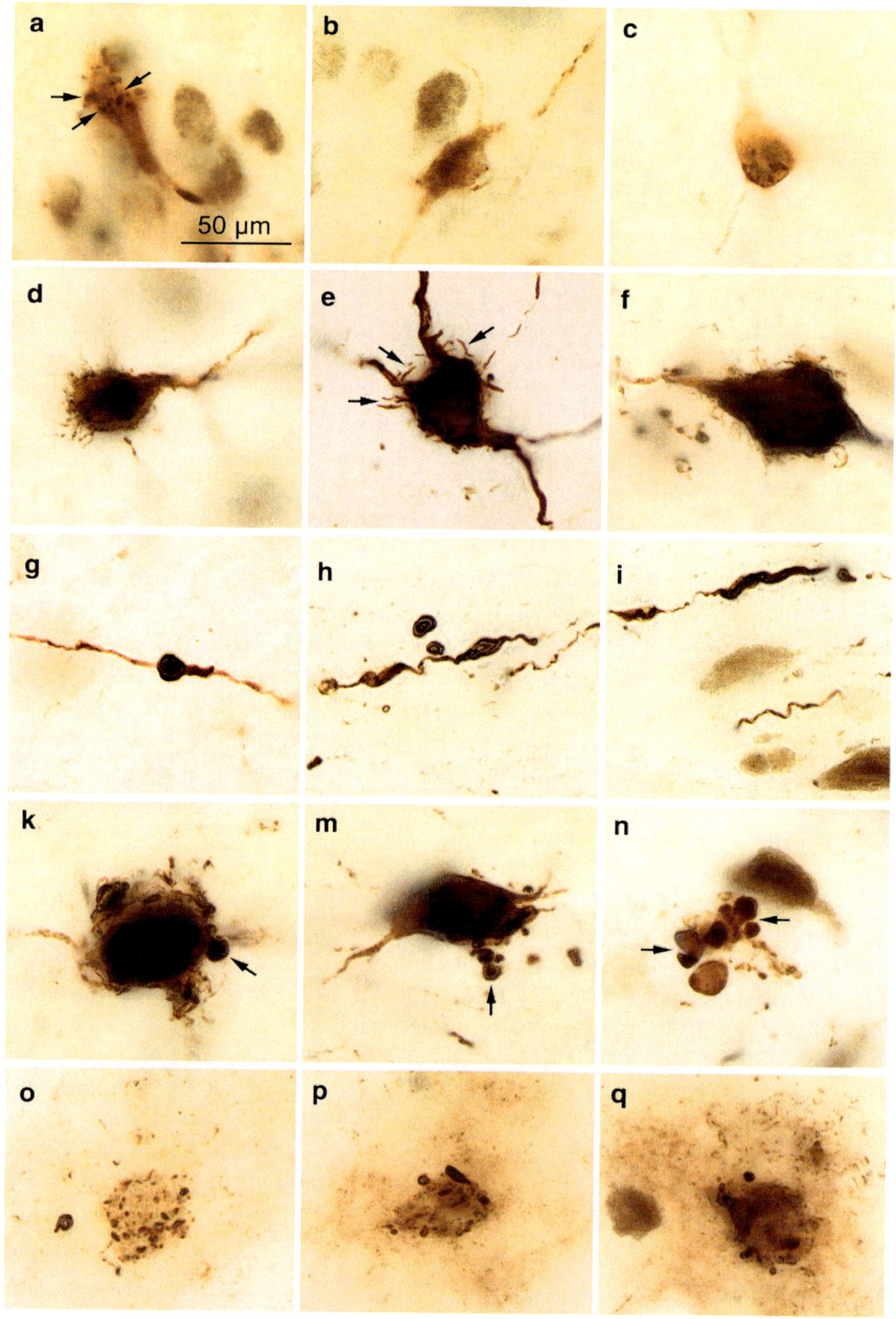

Fig. 5.2 AT8-immunoreactions in locus coeruleus neurons and astrocytes. (**a–c**) Droplet-like inclusions (*arrows*) in a 21-year-old male. (**d–f**) Short dendrite-like AT8-ir spikes (*arrows* in **e**) develop on the formerly smooth surface of the cell soma. (**g–i**) Varicosities in rarified dendrites that become fragmented and occasionally detached from the main branch (**h**). (**k–n**) Towards the end of these developments, the dendrites appear as radically truncated appendages in the form of

completely filled with uniformly weakly AT8-ir and with Gallyas-positive material. The neuronal somata no longer contain large deposits of neuromelanin granules (Fig. 5.3i). Severely altered remnants of dendritic processes with intensely AT8-ir enlargements can be found in the immediate vicinity (Fig. 5.2k–n). These findings lend support to the conjecture that the abnormal tau aggregates in astrocytic processes do not originate there but probably develop in response to the neuronal pathology. In more advanced disease stages (stages V and VI), AT8-ir astrocytes are no longer encountered anymore in the locus coeruleus (Fig. 5.3k).

5.3 Survival of Involved Neurons, Loss of Neuronal Function, and Degradation of Remnants After the Death of Involved Neurons

In general, projection neurons, including those vulnerable to the AD process, are remarkably sturdy because they possess diverse mechanisms for repairing transient damage, and they generally survive and continue to function as long as the individual lives. It is therefore not surprising that nerve cells do not undergo cell death during periods when they are producing large amounts of pretangle and tangle material (Bobinski et al. 1998: Morsch et al. 1999; Hof et al. 2003). Based on this fact, presently little understood mechanisms must exist whereby the cytoskeleton fulfills at least some of its functions, including those involving the axonal microtubuli. Thus, the ultimately insoluble and inert intra-axonal aggregates obviously do not fully prevent host axons from functioning, and the cell soma does not display reactive changes associated with axotomy, despite the presence of the space-consuming neurofibrillary lesions (Braak and Del Tredici 2011, 2012).

In this context, it is important to emphasize that the outcome of the AD-associated process is not primarily determined by global neuronal loss but, rather, it is the result of enormous numbers of surviving nerve cells with severely compromised or lost functionality (Stokin and Goldstein 2006; Merino-Serrais et al. 2013). Whether a modest decrease in the number of nerve cells or reduced functioning remains insignificant or becomes deleterious to brain functions also depends on the location of the involved neurons. The random distribution of a few impaired or dead cells probably is of less consequence, whereas a non-random distribution of approximately the same number of projection cells almost certainly will result in specific functional impairments. Notably, the topographic distribution

Fig. 5.2 (continued) shot-like pellets (*arrows*). (**o–q**) Abnormal and weakly AT8-ir astrocytic processes surround the remnants of dead dendrites and cell somata. Micrograph in (**d**) is from a 20-year-old female, in (**f**) from an 86-year-old female. Those in (**e**), (**g**), and (**n–q**) are from an 85-year-old male, (**h**) and (**i**) from a 76-year-old female, (**k**) from a 76-year-old male, and (**m**), (**p**), (**q**) from a 72-year-old male. 100 μm sections. Scale bar in (**a**) applies to (**b–q**)

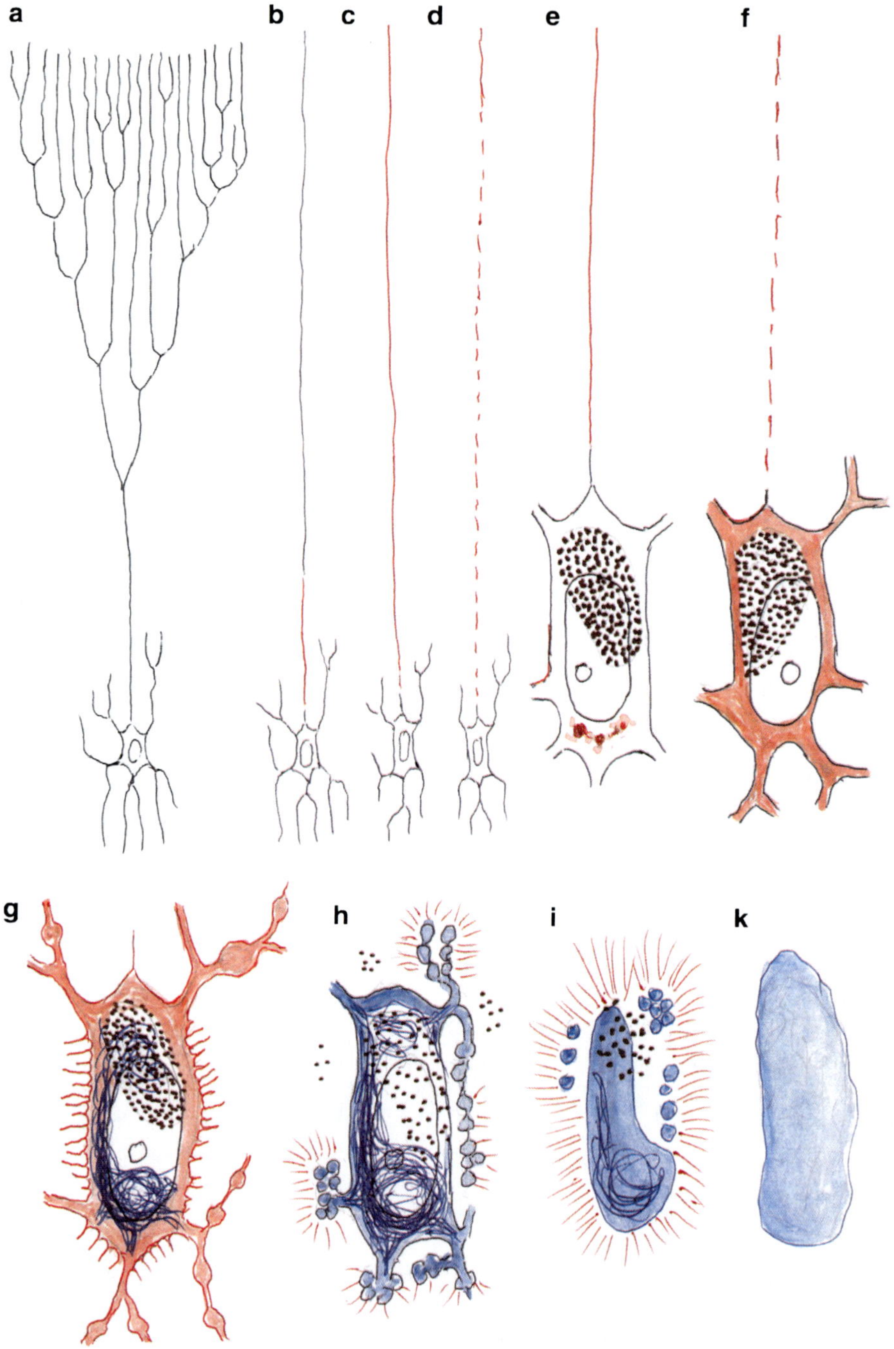

Fig. 5.3 Development of AD-associated lesions in neuromelanin-containing projection neurons of the locus coeruleus. (**a**) Locus coeruleus neuron projecting diffusely to the cerebral cortex. Note the widely branching axon (the length of the proximal axon is artificially shortened, so as not to exceed the permitted page space). (**b**) AT8-ir material is seen initially to fill the proximal axon in a

pattern of impaired nerve cells and the quantitatively limited neuronal loss in AD is strikingly non-random and, in most cases, bilaterally symmetrical.

Following the deaths of involved nerve cells, the abnormal material in axons ('tombstone axons,' as it were) is no longer visible. Obviously, then, both soluble and insoluble intra-axonal tau is re-absorbed within a relatively brief period of time. Furthermore, there is no light microscopically detectable indication that activated astroglial or microglial cells have migrated to the sites where widespread axonal loss has occurred, for instance, within the course of the perforant pathway (Streit et al. 2009, 2014).

The dense network of NTs frequently seen near newly-involved neurons and surrounding degenerating cortical pyramidal cells initially creates the impression that the dendritic trees of such cells remain intact. In fact, however, the network often is a non-functioning mass of *extra*-dendritic filaments (Fig. 7.2d, e), as is clearly evidenced by the increasing loss of their Gallyas positivity. Such extra-dendritic networks can be seen, chiefly, above the external cellular layer (pre-α) in the entorhinal region (Fig. 9.3a, b). Following the deaths of the cellular processes or their host somata (Fig. 9.3b), we think that the AD-associated pathological material in dendrites temporarily appears in the neuropil as "tombstone NTs" and—similar to the material in axons—is catabolized without leaving behind any remnants whatsoever (Ashford et al. 1998).

After cell death, somatic NFTs also lie free in the neuropil (Fig. 7.2g–i). The absence of a cell nucleus (when examined in sufficiently thick sections) makes it possible to identify these structures as tombstone tangles (Figs. 5.3k, 7.2g–l, and 9.4g, h). The material in tombstone tangles contains truncated tau proteins and varying combinations of other proteins, some of which may have become trapped

Fig. 5.3 (continued) uniform manner (stage a). (**c**) The abnormal tau material spreads gradually into more distal segments of the axon and does not reveal any immunonegative portions. (**d**) Directly after this phase, short AT8-ir axonal inclusions develop, all approximately the same length, and interspersed by immunonegative segments. The axon remains in this state for a period of time. (**e**) Following the changes within the axon, AT8-ir material also develops in the somatodendritic compartment of the cell, at first as droplets that rapidly become confluent. (**f**) This material fills the entire cell soma and dendrites without displacing the cell organelles (stage b). Here, the drawing depicts only the proximal dendritic stems. (**g**) Next, the dendrites rapidly develop varicosities as well as constrictions, and they contract. At the same time, new spike-like protrusions develop on the surface of the soma. Within the soma, the first particles of AT8-ir pathological material convert into argyrophilic filaments (here, depicted in *blue*). The very first filaments often originate near neuromelanin granule deposits. (**h**) The abnormal material in the somatodendritic compartment quickly forms neurofibrillary tangles and neuropil threads. At this stage, astrocytes can appear nearby dying nerve cells and near the deformed dendrites. The weakly AT8-ir processes of such astrocytes display a corona. (**i**) Following neuronal cell death, the nucleus is lost, but some of the neuromelanin granula and abnormal argyrophilic material can still be recognized in the form of droplets (dendritic remnants) and a tombstone tangle. These same remnants also can display a corona consisting of astrocytic processes. (**k**) Later on, even these processes perish, and a homogeneous filamentous mass is left over for a longer period of time, during which it is weakly AT8-ir and decreases its Gallyas-positivity

within NFTs during NFT formation. The fact that tombstone tangles are never observed in the absence of fresh AT8-ir pretangles or intraneuronal NFTs accounts for the absence of spontaneous remission in AD patients.

Clusters of lipofuscin or neuromelanin granules are likewise difficult to eliminate and, thus, they initially decorate the surfaces of tombstone tangles (Figs. 5.3i and 7.2g, h). Pigment granules, however, are among the first entities to become engulfed by astrocytes and thereafter gradually disappear from the tombstone tangles. The densely intertwined filamentous tombstone tangles slowly take on a looser aspect, display reduced AT8-reactivity, and subsequently forfeit their Gallyas-positivity (Fig. 7.2g–i). By contrast, their Campbell-Switzer staining profile is stronger than previously. This staining trait may be attributable to a gradually growing coating of Aβ protein (Munoz and Wang 1992). Tombstone tangles are demonstrable in the neuropil for years (Bobinski et al. 1998), although in cases where the AD process is far advanced even the tombstone tangles eventually disappear from the tissue and are replaced by a glial scar. Remarkably, a predilection site for such an event is, once again, the cellular islands residing in layer pre-α of the entorhinal cortex (Fig. 9.3g, h).

Chapter 6
Basic Organization of Territories That Become Sequentially Involved After Initial Involvement of Brainstem Nuclei with Diffuse Projections

6.1 The Cerebral Cortex

The cerebral cortex is the overarching superordinate structure of the human CNS. It is not a uniform entity but consists of two fundamentally different types of gray matter, i.e., the expansive and, for the most part, uniformly composed neocortex (Fig. 6.1a, yellow, light and deep orange, light and deep blue shading) and a small and heterogeneous allocortex (Fig. 6.1a, red shading). A transition region mediates between the two (Fig. 6.1a, green shading) (Braak 1980; Nieuwenhuys 1994; Amunts and Zilles 2001; Zilles and Amunts 2010). The neocortex is chiefly responsible for processing and planning the interactions with the external world. It receives abundant somatosensory, visual, and auditory data, and it influences, at the same time, somatomotor activity that impinges on the organism's environment (Fig. 6.1c). The neocortex, which takes up approximately 95 % of the total surface area of the human cerebral cortex, generally shows a six-layered organization, with the exception of a few regional variations.

Neocortical areas of the parietal, occipital, and temporal lobes are fundamentally divided into highly refined and maturely myelinated primary fields (core fields) that are responsible for the initial processing of incoming sensory data (Fig. 6.1a, deep blue). Each of the diverse sensory cores is encircled by a zone of somewhat less highly differentiated and only moderately myelinated unimodal first order sensory association areas (Fig. 6.1a, light blue). These first order areas, in turn, are interconnected with extensive but more simply organized and incompletely myelinated unimodal or heteromodal high order sensory association areas (Fig. 6.1a, yellow). The frontal lobe is similarly structured into a primary motor (core) field (Fig. 6.1a, deep orange) and premotor (belt) areas (Fig. 6.1a, light orange), the latter

H. Braak, K. Del Tredici, *Neuroanatomy and Pathology of Sporadic Alzheimer's Disease*, Advances in Anatomy, Embryology and Cell Biology 215,
DOI 10.1007/978-3-319-12679-1_6

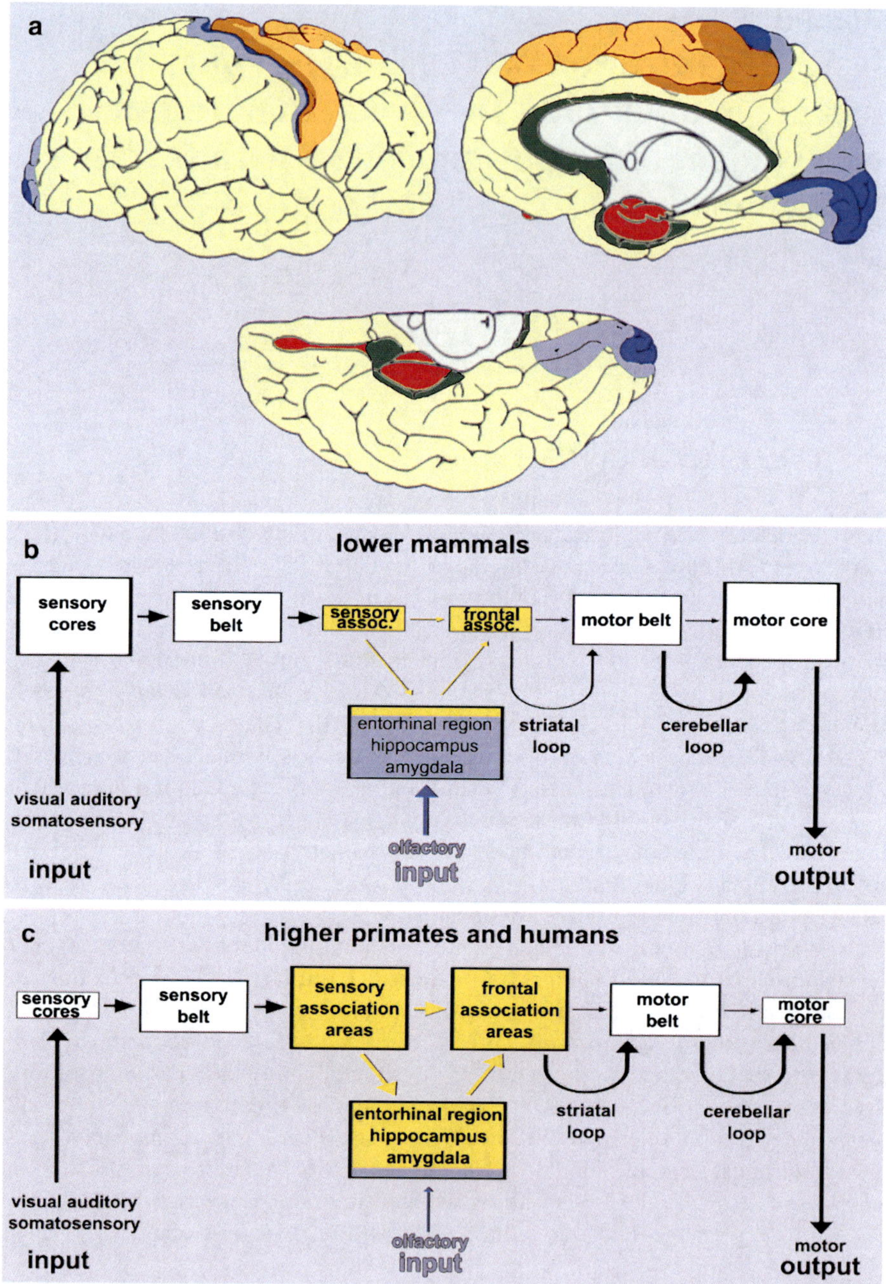

Fig. 6.1 Major subdivisions of the cerebral cortex. (**a**) The cerebral cortex consists of two basic types of gray matter: an extensive neocortex and a small allocortex (*red*). *Green* areas of the periallocortex and proneocortex mediate between the two. The allocortex is located chiefly in anteromedial portions of the temporal lobe. It includes the olfactory bulb and anterior olfactory nucleus, on the one hand, and limbic system centers, such as the hippocampal formation, presubiculum and entorhinal region, on the other. The subcortical amygdala is closely related. The neocortex is remarkably extensive in the human brain. The frontal, parietal, occipital, and

of which give way to the extensive high order frontal association areas, the prefrontal cortex (Fig. 6.1a, yellow). The enormous dimension of the prefrontal cortex, this highest executive instance in the CNS, is a distinct feature of the human brain.

By comparison, the heterogeneously composed allocortex is small (Fig. 6.1a, red). The number of layers varies from one region to another, ranging from simpler trilaminate structures (e.g., dentate fascia) to more complex structures with more than six layers (e.g., entorhinal region). The mammalian brain contains two allocortical cores: first, the olfactory bulb and areas related to it and, second, the hippocampal formation, presubiculum, and entorhinal region (Braak and Braak 1992a; Braak et al. 1996). In the microsmatic human brain, most structures devoted to olfactory processing are rudimentary (Fig. 6.1c) in comparison to the corresponding areas in macrosmatic mammals (Fig. 6.1b), whereas the hippocampal formation and regions related to it are highly developed. The hippocampus lacks direct sensory input and must rely, instead, on indirect input from both olfactory structures and the neocortex to obtain information about the world outside the individual (Fig. 6.1c).

Transition zones between the allocortex and neocortex include a zone of periallocortical areas allied with the allocortex in the narrowest sense and proneocortical areas leading to the mature neocortex (Fig. 6.1a, green). Because allocortical layers can transgress the limits of their own territories, the periallocortex often appears as an admixture of both allocortical and neocortical layers, whereas the proneocortex is usually dysgranular in character with poorly developed layers II and IV (Braak 1980).

6.2 The Amygdala

The amygdala is located in the anteromedial portion of the temporal lobe directly anterior to the hippocampal formation (Fig. 6.2). It chiefly consists of medial, cortical, and central subnuclei, supplemented by a voluminous basolateral complex that includes lateral, basal, and accessory basal subnuclei (Fig. 6.2) (Alheid 2003; Yilmazer-Hanke 2012; Bzdok et al. 2013). The amygdala is bilaterally and closely interconnected with cortical components of the limbic system: the entorhinal region, presubiculum, and hippocampal formation. Projections from olfactory

Fig. 6.1 (continued) temporal lobes each have a primary core area (*deep orange* and *deep blue*), a belt of first order association areas (*light orange* and *light blue*), and related high order association areas (*yellow*). (**b**) Macrosmatic mammals process olfactory input, in part, in the entorhinal region, which has only weak connectivities to the less well developed high order association areas of the neocortex. (**c**) By contrast, microsmatic primates and humans possess limited olfactory input and utilize the entorhinal region to process essentially neocortical information. The much larger high order association areas are connected to the 'newly' organized entorhinal region via the strong afferent and efferent trunks of the limbic circuit

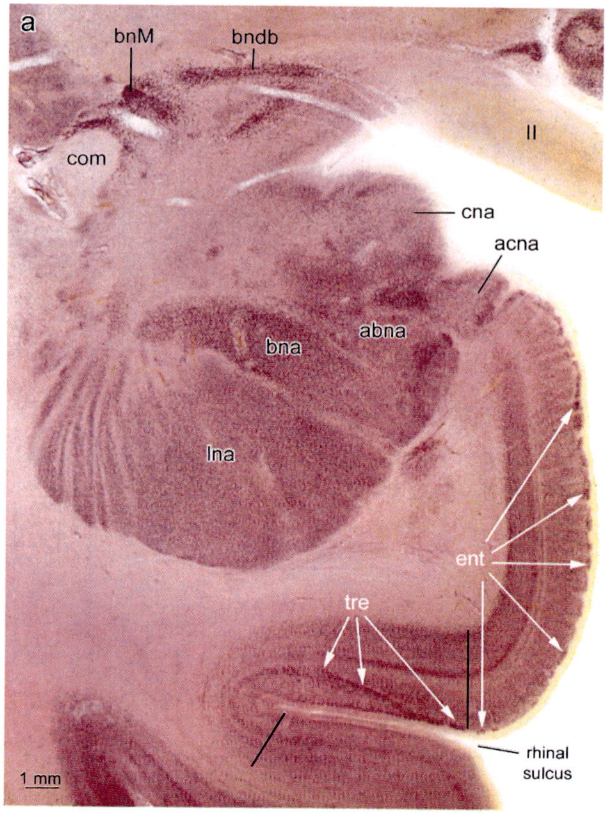

Fig. 6.2 Pigment-Nissl stained 200 μm section of the human entorhinal region cut at the latitude of amygdala. The entorhinal region (**ent**) covers almost totally the free surface of the parahippocampal gyrus, whereas the transentorhinal region (**tre**) is concealed deep in the rhinal sulcus. The intensely staining islands of pre-α (*white arrows*) coalesce in the vicinity of the rhinal sulcus and descend obliquely through the outer layers of the transentorhinal region the borders of which are indicated by *black lines*. The amygdala at this latitude shows cortical (**cna**) and accessory cortical subnuclei (**acna**). The large basolateral complex consists of three subnuclei, the accessory basal subnucleus (**abna**), basal subnucleus (**bna**), and lateral subnucleus (**lna**). *Abbreviations*: **acna**—accessory cortical subnucleus of the amygdala; **bndb**—bed nucleus of the diagonal band; **bnM**—basal nucleus of Meynert; **cna**—cortical subnucleus of the amygdala; **com**—anterior commissure; **II**—optic tract

areas terminate preferentially in cortical subnuclei, whereas the central subnucleus receives crucial interoceptive input from internal organs and regulates not only endocrine and autonomic functions but also learning, memory, and emotional reactions. Moreover, the central subnucleus controls all of the non-thalamic nuclei that generate diffuse projections to the cerebral cortex (Bohus et al. 1996; Nieuwenhuys 1996). The nuclei of the basolateral complex have bidirectional interconnectivities with neocortical high order association areas (Fig. 6.2). The lateral subnucleus is the major entry point for neocortical information. Dense projections from the basal and accessory basal subnuclei feed into the efferent trunk of the limbic circuit, which, in the human brain, heavily influences the prefrontal cortex via the ventral striatum, ventral pallidum, and dorsomedial nuclei of the thalamus (Fig. 6.10) (Heimer et al. 1991; Alheid 2003; Haber and Gdowski 2004).

6.3 The Entorhinal Region and the Presubiculum

The entorhinal region is the only association cortex of the hippocampal formation and extends over the ambient gyrus and anterior portions of the parahippocampal gyrus where small wartlike elevations (*verrucae hippocampi*) separated by shallow grooves mark the position of the entorhinal region and are visible to the naked eye (Fig. 6.4a) (Augustinack et al. 2012). The borders of the entorhinal cortex often coincide with the course of the rhinal sulcus anteriorly and the collateral sulcus posteriorly (Fig. 6.4a) (Hanke 1997). Presently, it is unknown whether and to what an extent variability of the entorhinal region reflects individual differences in personality or intellectural capacity (Amunts et al. 2005).

Notably, none of the entorhinal layers corresponds to a layer of the neocortex. To avoid confusion, a terminology distinguishing between external and internal principal strata (pre *versus* pri) is used throughout the text here (Figs. 6.3 and 9.1) (Braak and Braak 1992a). Large multipolar projection cells of layer pre-α are bunched together into characteristic cellular islands (Fig. 6.3), which occur nowhere else in the cerebral cortex. The external layers, pre-α, pre-β, and pre-γ give rise to the perforant pathway, which crosses the obliterated hippocampal fissure and terminates in the outer two-thirds of the dentate molecular layer, subiculum, and first sector of the Ammon's horn (CA 1) (Fig. 6.4c) (Witter 1993).

The presubiculum consists of cloud-like accumulations of small projection neurons supplemented by deep layers of the hippocampal subiculum, whereas the adjacent parasubiculum includes small neurons interfacing with a deep layer of the entorhinal region (Fig. 6.4b) (Kalus et al. 1989). Various versions of pathways subsumed under the name 'Papez circuit' arrive at the entorhinal region via the presubicular parvocellular layer (Fig. 6.10).

The transentorhinal region is situated along the lateral circumference of the entorhinal region deep in the rhinal sulcus (Figs. 6.2 and 6.4a, b) (Taylor and Probst 2008; Ding and van Hoesen 2010). The region is largest in the human brain and decreases markedly in size further down the primate scale (Braak and Braak 1992a). The transentorhinal region is easily identifiable by the gradual oblique descent of layer pre-α through the outer main stratum of the cortex (Fig. 6.2). Owing to this distinctive attribute, the transentorhinal region is one of the best characterized areas of the human cerebral cortex.

Together, the transentorhinal and entorhinal regions in higher primates and humans function as an interface between the neocortex and the hippocampal formation (Fig. 6.4c, d). Since virtually all of the data exchanged between the neocortex and hippocampus must pass through both regions (or else through the

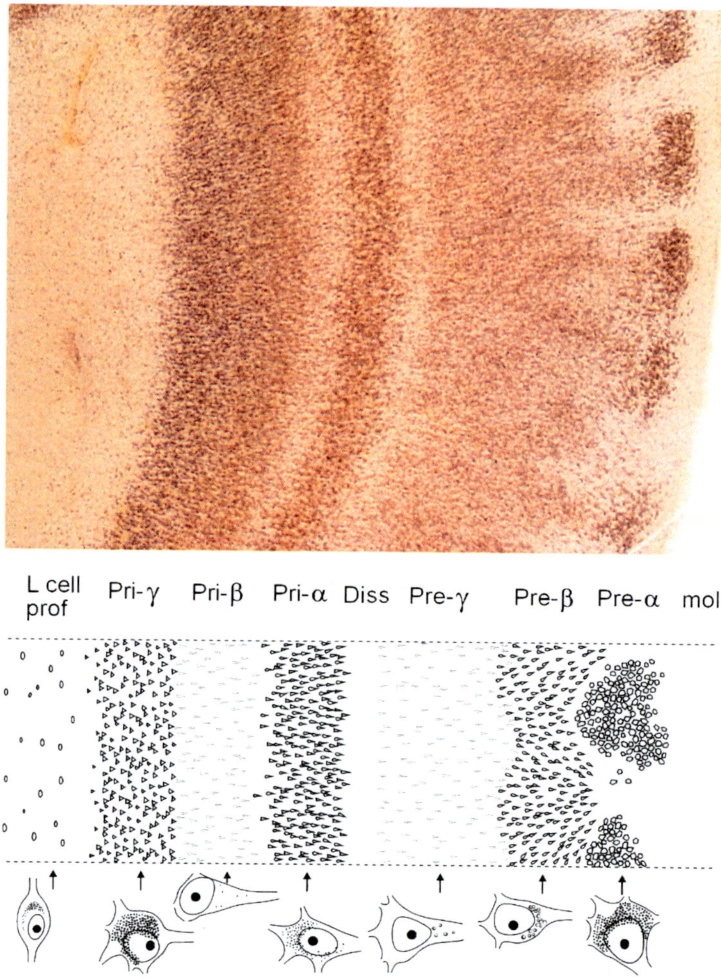

Fig. 6.3 Pigment-Nissl stained 200 μm section of the human entorhinal cortex and a detailed scheme of lamination as well as characteristic nerve cells types in nine layers based on pigmentation (lipofuscin granule) patterns. Drawing shows the cellular layers located in the outer and inner main *strata* (Pre and Pri). *Abbreviations*: **mol**—molecular layer; **Pre-α**—layer pre-α, **Pre-β**—layer pre-β, **Pre-γ**—layer pre-γ (cellular layers of the outer main stratum—external principal stratum); **Diss**—Lamina dissecans; **Pri-α**—layer pri-α, **Pri-β**—layer pri-β, **Pri-γ**—layer pri-γ (cellular layers of the inner main stratum—internal principal stratum); **L cell prof**—*Lamina cellularis profunda*. Diagram adapted with permission from H Braak and E Braak, Temporal sequence of Alzheimer's disease-related pathology. Cerebral Cortex 1999;14:475–512

lateral subnucleus of the amygdala), bilateral damage to the transentorhinal and entorhinal regions and the amygdala impairs this dataflow and ultimately interferes with higher cortical capabilities (Fig. 6.10).

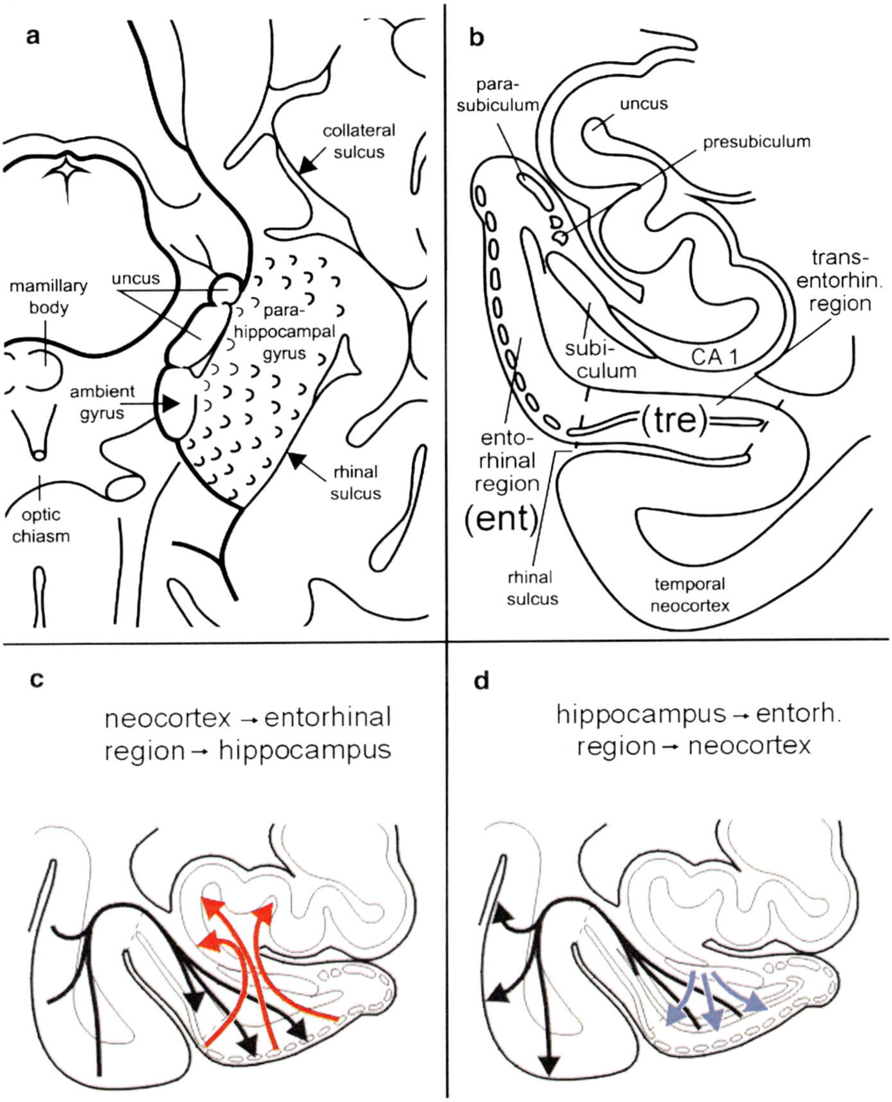

Fig. 6.4 Macroscopic topography of the human entorhinal cortex and schemata of major entorhinal region afferent and efferent projections. (**a**) Diagram showing the topographical relationships of the entorhinal region, which extends over anteromedial portions of the temporal lobe. The surface of the entorhinal cortex (parahippocampal gyrus) exhibits wart-like elevations, the *verrucae hippocampi*. Note the uncus of the hippocampus, which reaches the free surface of the brain. (**b**) Frontal section through the uncal portion of the hippocampus, the adjoining parahippocampal gyrus and occipitotemporal gyrus. (**c**, **d**) Diagrams of the major entorhinal connectivities. (**c**) Neocortical input (*black arrows*) together with input from limbic circuits. The borders of the entorhinal and transentorhinal regions are marked by a dashed line. Converges upon the

Given the unique status of the entorhinal region in the human brain, the region is shown here in sections cut tangentially to the cortical surface, so as to display the patterns of the approximately 250 cellular islands in the superficial layer pre-α (Fig. 6.5a–d). The pattern of each individual is unique (compare Fig. 6.5c with d): in some cases, the islands are large but few (Fig. 6.5d), in others numerous small islands are present (Fig. 6.5c). In some individuals, a large proportion of the cellular islands in the pre-α layer have a reticulate pattern, in others they are detached. The right and left entorhinal regions of each person are similar, like the fingerprints of both hands (Fig. 6.5a right, b left). Remarkably, the entorhinal cortex—and, above all, its superficial layer of cellular islands, which transfers crucial information between the neocortex and the hippocampal formation—is highly susceptible to early-appearing and severe pathological changes in the AD process (Braak and Braak 1991a, 1999).

6.4 The Hippocampal Formation

The hippocampus is almost completely hidden in the depths of the temporal lobe and gradually decreases in size en-route to the splenium of the corpus callosum. Three major units can be distinguished: the dentate fascia, the Ammon's horn (*cornu Ammonis*) with its four sectors (CA 1–CA 4), and the subiculum (Fig. 6.6). In the neocortex, layers that preferentially receive afferent fibers and those generating efferent projections are arranged one on top of the other. However, in the hippocampal formation, the layers lie side by side. The dentate fascia is a kind of hippocampal granular cortex that is specialized at receiving a number of different afferents. It consists chiefly of small projection cells that functionally correspond to the granule cells of the fourth neocortical layer (spiny stellate cells). Both the Ammon's horn and the subiculum, by contrast, are dominated by large pyramidal cells that furnish the output pathways and functionally correspond to neocortical layers III and V (Braak et al. 1996; Gloor 1997; Insausti and Amaral 2012).

The glutamatergic perforant path originates in the entorhinal region and provides the most important excitatory input to the hippocampus, where it preferentially terminates within the outer two-thirds of the molecular layer of the dentate fascia

---

Fig. 6.4 (continued) transentorhinal region and layer pre-α of the entorhinal region. From there, it is transferred through the perforant pathway (*red arrows*) heading to the hippocampal formation. (**d**) A dense back-projection from the subiculum of the hippocampal formation terminates in the entorhinal deep layer pri-α (*red arrows*) and is then transferred back to the neocortex (*black arrows*). Thus, the entorhinal and transentorhinal regions serve predominantly as an interface between the neocortex and hippocampal formation. *Abbreviations*: **CA 1**—*cornu Ammonis* (first sector of the Ammon's horn); **ent**—entorhinal region; **tre**—transentorhinal region. Micrographs (**a**) and (**b**) are adapted with permission from H Braak et al., Acta Neuropathol 2006;112:389–404

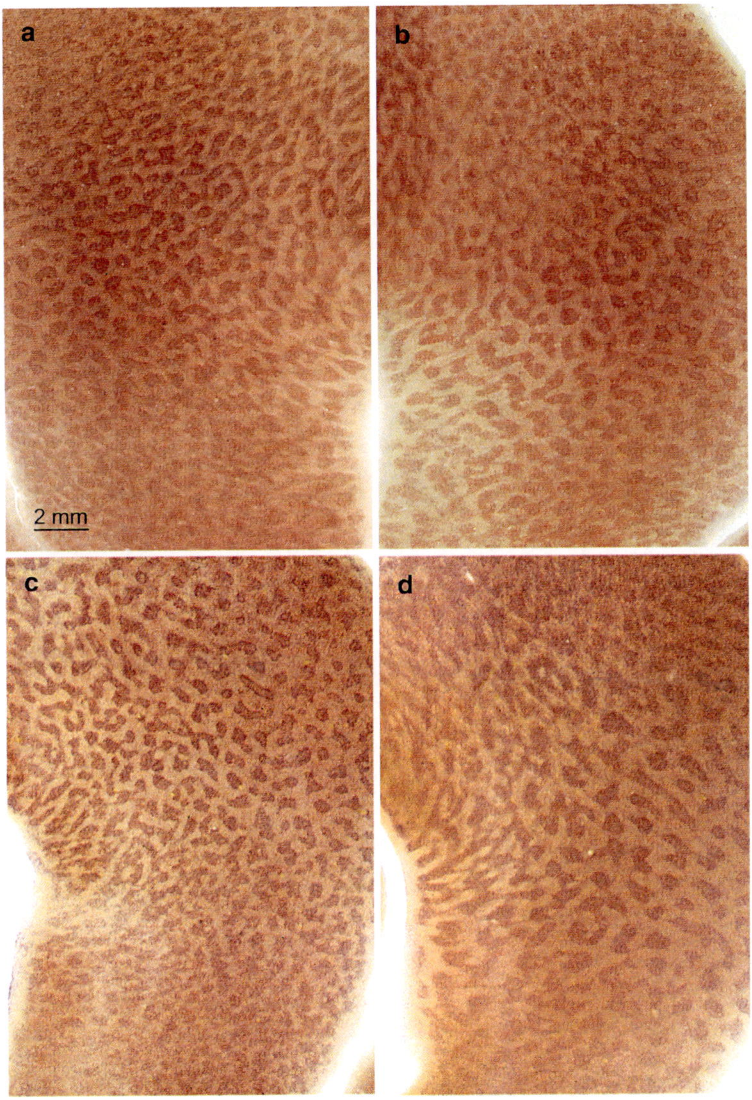

Fig. 6.5 Pigment-Nissl stained 200 μm sections cut tangentially to the surface of the entorhinal cortex from individual brains. In all sections, the highly unusual cellular islands of layer pre-α are visible. Whereas the patterns in layer pre-α of the left and right entorhinal cortex of a single individual resemble each other very closely (**a**, **b**), the unique patterns of different individuals make it possible to distinguish these from each other (**c**, **d**). Reproduced with permission from H Braak and E Braak, Staging of Alzheimer's disease-related neurofibrillary changes. Neurobiol Aging 1995;16:271–284. Scale bar in (**a**) also applies to micrographs in (**b**–**d**)

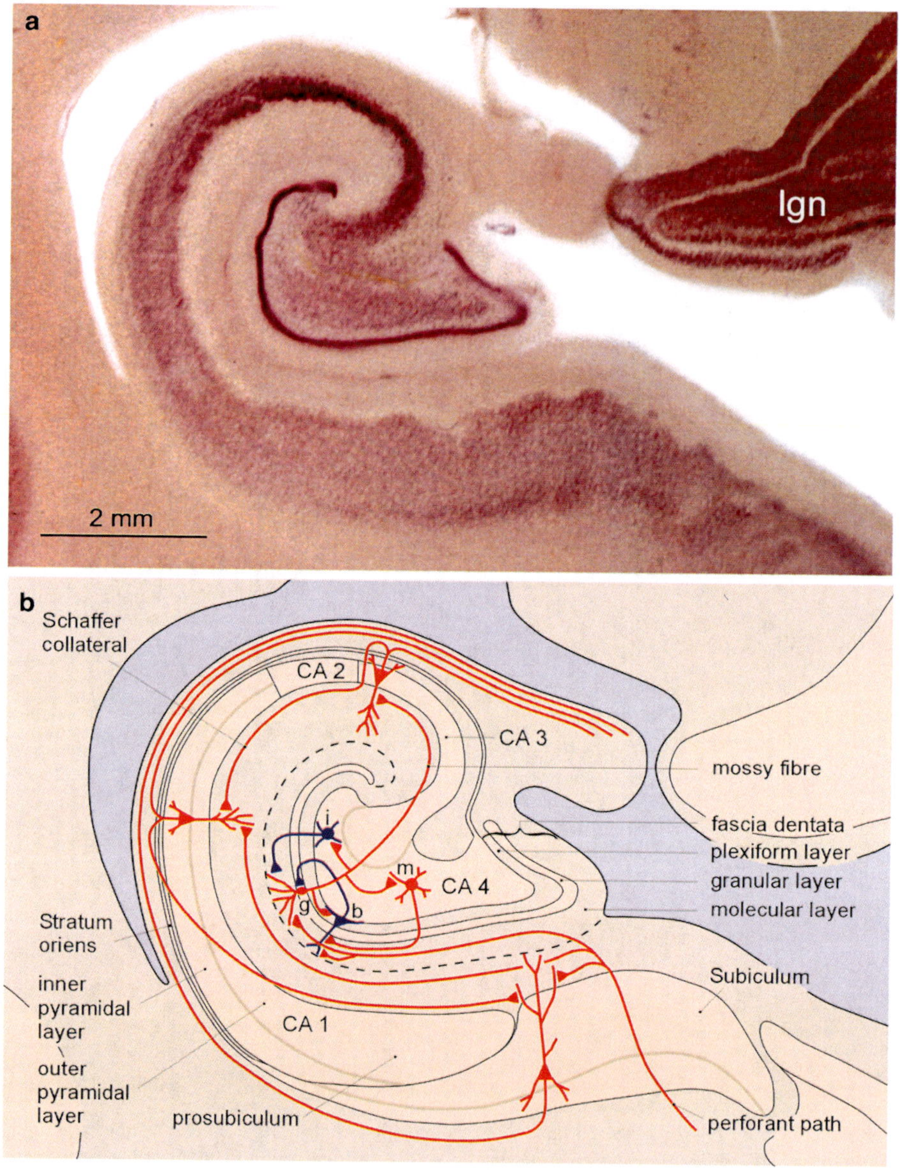

Fig. 6.6 Pigment-Nissl stained 200 μm section of the Ammon's horn and a schematic template. (**a**) Frontal section of the hippocampus at the level of the lateral geniculate nucleus (**lgn**) (aldehydefuchsin and *Darrow red*). (**b**) Schematic drawing of the sectors of the Ammon's horn (CA 1–CA 4) layers, cell types, and principal connections. *Abbreviations*: **b**—basket cell type (pyramidal forms in the dentate fascia); **g**—granule cell in the dentate gyrus; **i**—interneuron in CA 4; **m**—mossy cell in CA 4; **CA 1–4**—sectors 1–4 of the Ammon's horn. Reproduced with permission from H Braak et al., Functional anatomy of human hippocampal formation and related structures. J Child Neurol 1996;11:265–275

(Fig. 6.6b, *fascia dentata*) (Insausti and Amaral 2012). Other important afferents originate from the cholinergic magnocellular nuclei of the basal forebrain. Fibers from these nuclei and association fibers from CA 3/CA 4 neurons branch off within the inner third of the dentate molecular layer. The granule cells of the dentate fascia (Fig. 6.6b, g) give rise to mossy fibers characterized by unusually large synaptic boutons. The mossy fibers profusely collateralize and terminate at dendritic excrescences of projection neurons in CA 3/CA 4. The zone of the terminating mossy fibers contains astrocytes with deposits of lipofuscin granules that do not occur anywhere else in the human brain (Braak et al. 1996). The glutamatergic projection neurons of CA 4 (Fig. 6.6b, mossy cells—**m**) give rise to abundant association fibers that establish a recurrent excitatory circuit with the dentate fascia. The pyramidal cells of CA 3 give rise to the Schaffer collaterals terminating within the stratum oriens and stratum radiatum of CA 1. CA 1 pyramidal cells, in turn, project to the subiculum (Fig. 6.6b).

The relative size of CA 1 increases considerably ascending along the primate scale (Stephan 1983). In humans, CA 1 consists of a superficial and deep pyramidal cell layer (Figs. 6.6a, b and 6.7a, b). A wedge-shaped portion of the superficial pyramidal layer overlaps the subiculum and is referred to as the prosubiculum (Fig. 6.6b). The enigmatic sector CA 2 stands out owing to its tightly bunched and voluminous pyramidal cells (Fig. 6.6a, b). Even today, little is known about the specific connections and functional significance of CA 2. The sector has minimal input from the mossy fibers (dentate fascia) and Schaffer collaterals (CA 3). Some specialized input probably originates from the hypothalamic tuberomamillary nucleus. The subiculum appears on cross-sections as a wing-shaped structure (Fig. 6.6b). It consists of a molecular layer, extended outer and inner pyramidal cell layers, and a deep layer originating in the entorhinal region. The subicular pyramidal cells give rise to a long apical dendrite which, in adults, contains a spindle-shaped lipofuscin deposit—a phenomenon that does not occur anywhere else in the cortex (Fig. 6.7c, d) (Braak et al. 1996). Hippocampal output is mainly generated in the subiculum. Efferent projections head toward the presubiculum and entorhinal region as well as toward the amygdala and ventral striatum (Braak et al. 1996; Insausti and Amaral 2012).

In summary, the amygdala and limbic allocortical regions are optimally positioned to extract from the enormous flow of incoming exteroceptive and interoceptive data the crucial information needed to produce an appropriate response to any given set of circumstances (Figs. 6.9b and 6.10).

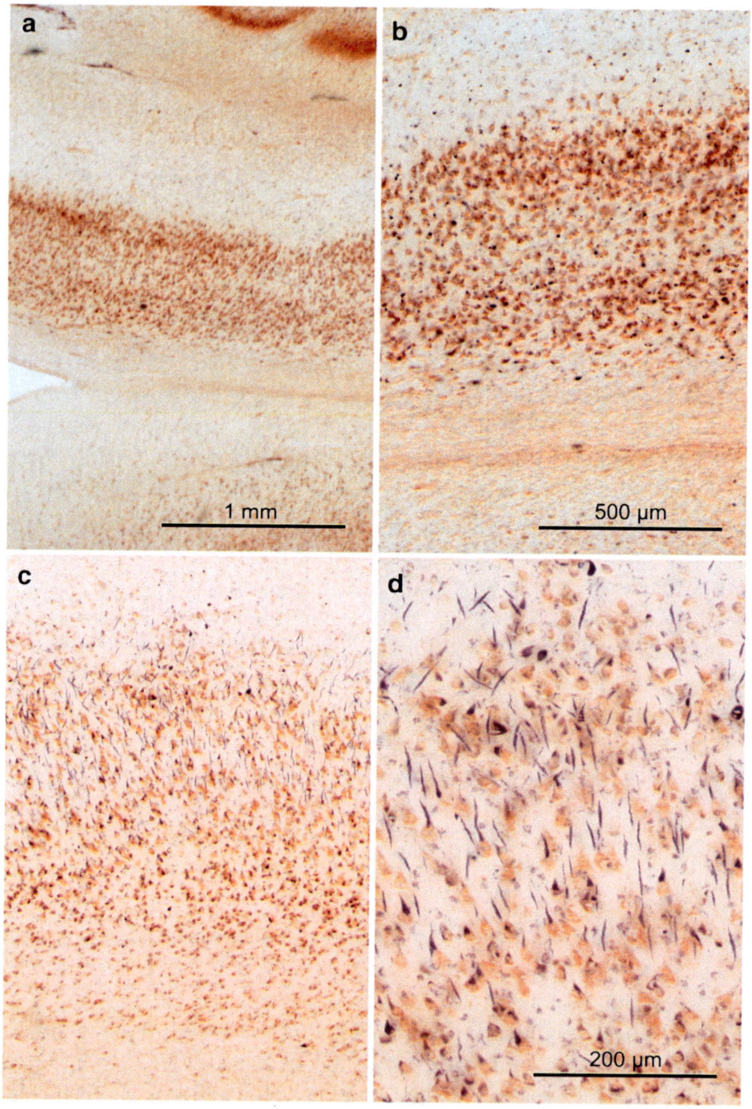

Fig. 6.7 Pigment-Nissl stained 200 µm frontal sections of the first sector of the Ammon's horn and subiculum. (**a**, **b**) Outer and inner pyramidal layers of CA 1. The inner layer overlaps the wedge-shaped subiculum and forms the prosubiculum. (**c**, **d**) Distinguishing features of the subiculum are pyramidal cells that contain within their apical dendrites a noticeably spindle-shaped deposit of lipofuscin granules

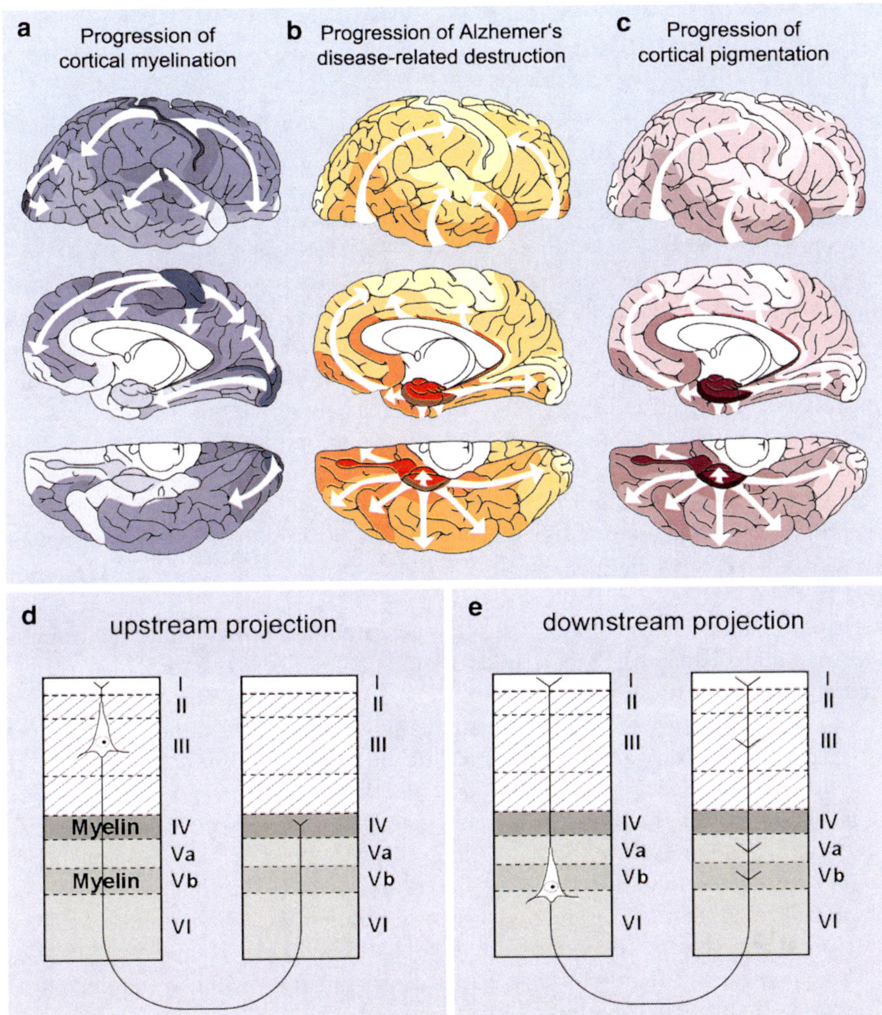

Fig. 6.8 Sequential progression of cortical myelination (**a**), Alzheimer's disease-associated tau pathology (**b**), and pigmentation patterns (**c**). (**a**) Cortical myelination begins in primary neocortical areas and progresses via secondary areas to related high order association fields (indicated by shades of *blue*). (**b**) AD-related cortical lesions begin in the transentorhinal region of the temporal lobe. From there, the pathology extends into adjoining high order association areas, eventually reaching the primary areas of the neocortex (indicated by shades of *red/yellow*). (**c**) The sequential pattern of cortical pigmentation closely resembles that of the AD-related destruction (indicated by *violet* shading). Note how the sequence of AD-associated neurofibrillary pathology inversely recapitulates the sequence of cortical myelination. (**d**) Upstream projection. Projections from primary areas to first order and/or high order association areas of the neocortex and from pyramidal cells of layer III predominantly terminate in layer IV. (**e**) Downstream projection. Projections from high order association fields and/or first order association areas directed to primary areas of the neocortex originate from subgranular pyramidal cells (layers V/VI) and terminate in layer I and other layers, virtually sparing layer IV

6.5 Cortical Gradients in Differentiation, Myelination, and Pigmentation

Cortical differentiation increases gradually when one draws an imaginary line from the proneocortex through high order and first order association areas into the most highly refined primary fields of the neocortex. This architectonic hierarchy is reflected by an equally hierarchical arrangement of the major pathways interconnecting the basic territories (Braak 1980; Mesulam 1998).

In lower mammals, the neocortex is small and consists mainly of primary fields, whereas the association areas are hardly developed (Fig. 6.1b). With phylogenetic advancement, however, the association areas expand enormously and the primary fields become subject to an ever greater degree of internal differentiation and functional sophistication (Fig. 6.1c). A good example of this tendency is the primary visual field. During its phylogenetic evolution, the data processing complexity in this field increases considerably, and this development is partially reflected by the increased number of cortical layers there. The complexity of the first order association areas assigns them a middle position between two extremes, i.e., the primary fields and the sensory high order association areas and prefrontal fields of the neocortex.

The end result of this trend is that the neocortex in humans, more than in any other mammal, displays the greatest differential gradient between the highly refined primary fields and the relatively simply constructed, but expansive, high order sensory association areas and prefrontal regions. These maturational differences among cortical regions are also reflected in the basic principles that govern the formation of myelin and lipofuscin deposits within the human cortex (Fig. 6.8a, c). Prenatal cortical myelination commences late, it progresses in a predetermined sequence, and it persists well into adulthood (Fig. 6.8a). Whereas components of the spinal cord and brain stem that are necessary for survival myelinate early, the cerebral cortex in general and its high order association areas in particular undergo a late-onset and very prolonged myelination process (van der Knaap et al. 1991).

This maturational hierarchy also has a direct bearing on the development of intracortical networks of preferentially tangentially aligned myelinated fibers. In lower mammals, a deep plexus in layer Vb (the inner line of Baillarger) is initially found, whereas in higher mammals and in primates a second myelin stripe appears in layer IV (the outer line of Baillarger). Moreover, in the human brain, an additional loosely arranged plexus in layer III (stripe of Kaes Bechterew) can be made out in some regions. As phylogenetic progress continues, this plexus has the potential to separate layer II from layer III and, in so doing, to add a third myelin stripe to the structure of the neocortex.

In early childhood, initial traces of myelin appear in the neocortical primary motor and primary sensory fields. From there, myelination goes on to include the first order sensory association areas and premotor fields and, ever so gradually, the high order sensory association areas and prefrontal fields (Fig. 6.8a). The end result is an exceptionally dense myelination (*typus dives*) of the primary fields in the adult. With increasing distance from these fields, a decrease in the average myelin content occurs, and the areas furthest removed from the primary areas, i.e., the temporal high order processing areas close to the allocortex, display a remarkably poor degree of myelination (*typus pauper*). This regional gradation is important for facilitating the distinction of architectonic units in the human brain. The myelin contents of neo- and allocortical areas cannot be directly compared with one another owing to major structural differences between the two. However, it is important to note that the allocortical cell layers are, for the most part, relatively sparsely myelinated and that the limbic fiber tracts, such as the fornix and the perforant pathway, commence and conclude myelination late (Vogt and Vogt 1919; Braak 1980; van der Knaap et al. 1991; Shaw et al. 2008).

The mean density of cortical intraneuronal lipofuscin deposition bears a puzzling likeness to the inverse process of cortical myelination (Fig. 6.8c) (Braak 1980). Cortical lipofuscin granules first become apparent in early adulthood and slowly increase in number. Richly myelinated cortical areas in the brain of the human adult, such as the primary motor field and the cortical layers fortified with a thick plexus of myelinated fibers (i.e., neocortical layers IV-Vb), generally are sparsely pigmented and light in appearance (*typus clarus*). By contrast, the thinly myelinated basal temporal neocortex and the transentorhinal region have a dark appearance (*typus obscurus*) and are among the most richly pigmented territories of the cortex (Fig. 6.8c) (Braak 1980). The projection cells in the superficial entorhinal cellular layer pre-α generating the perforant pathway are particularly rich in lipofuscin granules (Fig. 6.3).

6.6 Interconnecting Pathways

The overview of the human cerebral cortex and related subcortical regions presented above provides a partial impression of the most essential interconnecting pathways. Pyramidal cells of supragranular layers (II-III) of primary sensory (core) fields give rise to short *upstream* projections that preferably reach layer IV of first order sensory (belt) areas. The same type of supragranular pyramidal cells sends long and sparsely myelinated axons to layer IV of the adjoining high order sensory association areas (Figs. 6.8d and 6.9a) (Rockland and Pandya 1979; Barbas 2007). In this manner, visual, somatosensory, and auditory data proceed sequentially from the core fields through the surrounding belt areas to the corresponding high order sensory association areas. From there, the data are conveyed via long and poorly myelinated ipsilateral corticocortical projections to layer IV of the prefrontal cortical areas (Fig. 6.9a). Layer IV consists chiefly of small pyramidal cells (spiny stellate cells) with a short axon. These neurons are specialized at receiving information effectively and distribute data radially to the remaining cell layers (Bannister 2005).

Short *downstream or return* pathways from the highest executive instance in the CNS, the prefrontal cortex, are provided by ipsilateral corticocortical projections

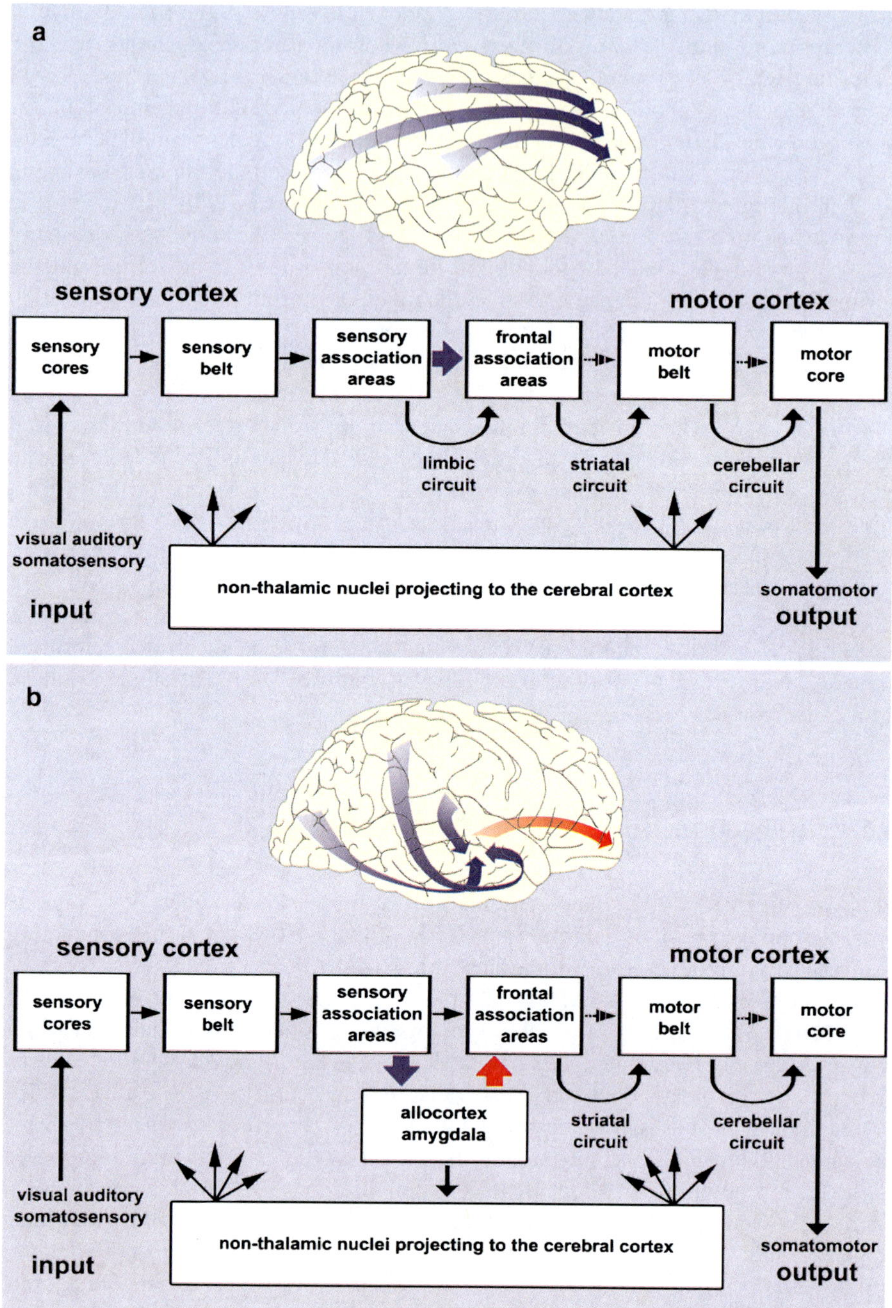

Fig. 6.9 Major projections between different regions of the cerebral cortex that receive diffuse input from non-thalamic subcortical nuclei. (**a**) Visual, auditory, and somatosensory information proceeds through the respective primary and first order association fields of the neocortex to a variety of high order association areas, whence it is transported via long and only moderately myelinated corticocortical pathways to the prefrontal association areas (*blue arrows*). Tracts

that preferably originate in infragranular layers (V-VI) and end chiefly in the molecular layer of their target areas (Fig. 6.8e), supplemented by targets in layers II, III, V and VI, while layer IV has only sparse contacts (Barbas 2007). The downstream pathways project to the cortex in a far less focused and effective manner than the upstream projections to layer IV. In this respect, the return pathways resemble the diffuse projections from the non-thalamic nuclei (Chap. 3), whereas the upstream projections are comparable to precise thalamocortical projections. The return pathways ultimately transmit data from the prefrontal fields via the premotor zones to the primary motor (core) field, which is the major gateway for relaying motor programs to neurons in the lower brainstem and spinal cord premotor and motoneurons (Fig. 6.9a, small dashed arrows).

The downstream dataflow lacks precision. However, parallel circuits direct most of the data into the striatal and cerebellar circuits, which integrate the basal ganglia, thalamus, many lower brainstem nuclei, the cerebellum, and the spinal cord into the regulation of cortical output (Figs. 6.1b, c and 6.9a) (Heimer et al. 1991; Alheid 2003; Haber and Gdowski 2004; Petrides and Pandya 2004; Heimer and van Hoesen 2006; DeLong and Wichmann 2007).

Limbic circuit components (the entorhinal region, presubiculum, hippocampal formation, and amygdala) intervene in this dataflow at a critical point, namely, where exteroceptive data is conveyed from the high order sensory association areas to the prefrontal cortex (Figs. 6.9b and 6.10). Because the neocortex is continuously flooded by non-essential information, the data valuable for the individual need to be continually filtered within the neocortex, until only the most essential contents reach the limbic circuit centers. To make this possible, the dataflow is split up: data from sensory high order association areas leaves the mainstream, converges through multiple neocortical relay stations and is siphoned (1) via the transentorhinal and entorhinal regions to the hippocampal formation and (2) via the lateral subnucleus of the amygdala to other components of this nuclear complex and points beyond (Fig. 6.10).

Limbic circuit output structures (the subiculum of the hippocampal formation and the basal and accessory basal subnuclei of the amygdala) generate dense projections to the accumbens nucleus and ventral striatum. These projections are supplemented by others from the midline nuclei of the thalamus (Fig. 6.10). From the ventral striatum, limbic circuit output is conducted via the ventral pallidum and the magnocellular mediodorsal thalamic nuclei to medial and orbitofrontal portions

Fig. 6.9 (continued) originating from this highest organizational level of the brain transmit the data back to the primary fields of the frontal lobe. Downstream connections are indicated by *dashed arrows*. The striatal and cerebellar circuits provide the major routes for this transport. (**b**) Select data en-route from the sensory association cortex to the prefrontal cortex diverge and eventually converge upon the transentorhinal/entorhinal regions and the amygdala, thereby constituting the afferent arm of the limbic circuit (*blue arrows*). In the human brain, the neocortex provides the most important input to the limbic system. Projections from the entorhinal region, amygdala, and hippocampus constitute the efferent arm of the limbic circuit (*red arrows*), which reaches the prefrontal cortex via the ventral striatum, ventral pallidum, and mediodorsal thalamus

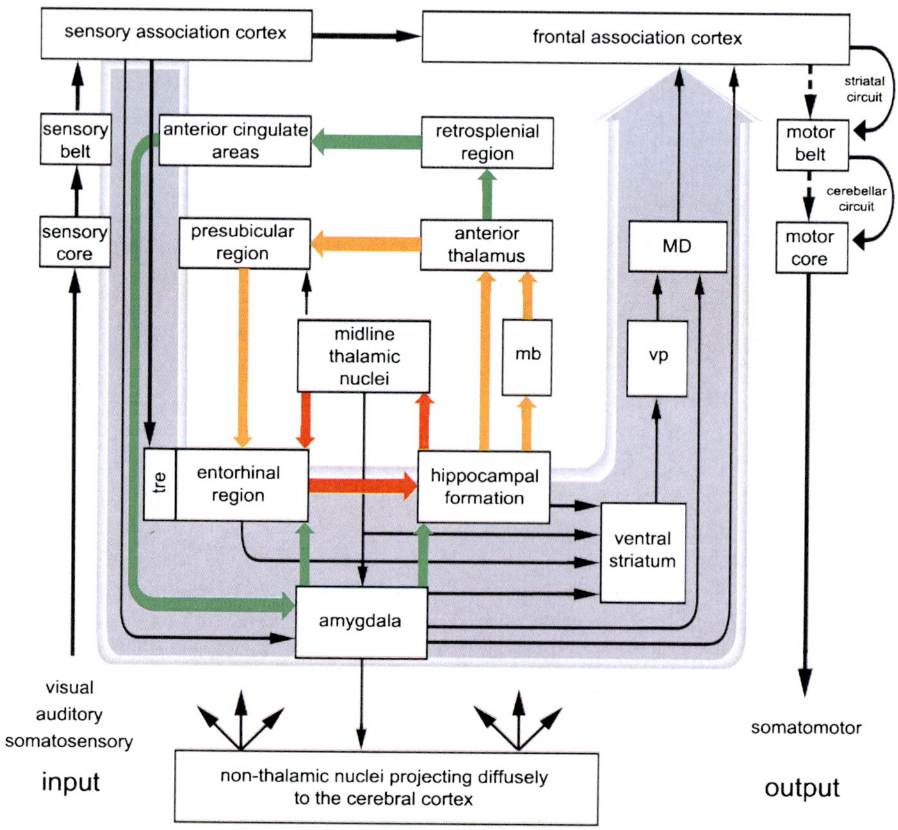

Fig. 6.10 Diagram depicting the limbic circuit in greater detail. The *gray arrow* emphasizes the circuit's strategic position between neocortical sensory association areas and the prefrontal cortex. The hippocampal formation, entorhinal region, and amygdala are densely interconnected. Together, they are the predominant organizational structures of the limbic system. Important limbic circuits representing different parts of the 'Papez circuit' are indicated by different color coding: *Red arrows*—reuniens circuit, afferent and efferent projections from and to the midline thalamic nuclei establish short thalamo-allocortical circuits. *Orange arrows*—this circuit is supplemented by others, including the anterior thalamus. One circuit is formed by projections to the presubicular region. *Green arrows*—a more extended circuit includes both the retrosplenial region and anterior cingulate areas as well as the amygdala. *Abbreviations*: **mb**—mamillary body; **MD**—mediodorsal thalamic nuclei; **tre**—transentorhinal region; **vp**—ventral pallidum

of the prefrontal neocortex. The basal and accessory basal subnuclei of the amygdala generate supplementary projections to the mediodorsal thalamic nuclei and prefrontal cortex. Fronto-amygdalar projections reciprocate these connections (see Fig. 6.10 for the efferent arm of the limbic circuit) (Braak et al. 1996; Gloor 1997; Holstege et al. 2004; Suzuki and Amaral 2004; Heimer and van Hoesen 2006; Groenewegen and Trimble 2007; Insausti and Amaral 2012).

The limbic circuit makes the neocortex the leading source of information for the human limbic system. During evolution from macrosmatic mammals to microsmatic higher primates and humans, a remarkable expansion of the neocortex accompanied by a thorough internal reorganization of its interconnectivities with the centers of the limbic circuit takes place (compare Fig. 6.1c with b). The major vestige of this process is an enormous increase in size and sophistication of those limbic circuit centers that predominantly receive neocortical input and generate neocortical output (limbic circuit, Fig. 6.10). The increase of neocortex-dependent transentorhinal and entorhinal fields occurs at the expense of the previously predominant territories involved in processing olfactory data (compare Fig. 6.1c with b).

The presubiculum and portions of the thalamus, basal ganglia, and proneocortex are also limbic circuit components (Fig. 6.10). They receive indirect somatosensory and viscerosensory input and project to systems subserving both somatomotor, visceromotor, and endocrine functions (Fig. 6.10). They not only control virtually all subcortical and cortical centers that regulate endocrine and autonomic functions but also are integral to the maintenance of learning, memory, and emotional equilibrium. At the same time they influence somatomotor activity because their input to the prefrontal cortex causes an individual's motor behavior to reflect his or her emotional state of mind. As custodians of memory and learning, the centers of the limbic circuit act as a neuronal bridge linking the external and internal worlds (Heimer et al. 1991; Mesulam 1998; Squire and Schacter 2002; Haber and Gdowski 2004; Heimer and van Hoesen 2006).

The limbic centers and the neocortex of the healthy human brain function harmoniously. The neocortex specializes in precise and initially unfiltered analyses of exteroceptive data. However, if data is to be retained in memory or categorized as significant, it has to be sorted out of the large quantities of incoming non-essential data. Cooperation between the neocortex and the centers of the limbic circuit makes this discriminatory function possible and is especially significant for understanding the consequences of the brain destruction that develops during the AD process.

Chapter 7
The Pattern of Cortical Lesions in Preclinical Stages

7.1 Stages 1a and 1b: Development of Inclusions in Axons and of Pretangle Material in Transentorhinal Pyramidal Cells

As the pathological process progresses and secondary to the subcortical lesions described above, abnormal tau aggregates appear for the first time in the cerebral cortex generally in the transentorhinal region. AT8-ir lesions are present in neuronal processes, most probably in terminal portions of axons, and we designate such cases as having "cortical stage 1a" tau pathology (Table 2.1; Figs. 7.1 and 7.2). The predominantly radially aligned thread-like cellular processes are easier to see in unconventionally thick sections (Fig. 7.1a, b). A thin network of finely shaped and weakly AT8-ir neuronal processes (axons) together with more intensely immunostained dots develops. Sometimes, however, it is possible to follow several of the processes over longer distances, and they often end with a droplet-like swelling (Figs. 7.1c, d and 7.2a). All stage 1a cases show the presence of subcortical tau lesions resembling those found in stages a–c (see Sects. 5.1 and 5.2).

During cortical stage 1b (Table 2.1, AT8-ir inclusions in cortical pyramidal cells), the stage 1a lesions are accompanied by AT8-ir transentorhinal pyramidal cells that closely resemble the multipolar neurons of the superficial entorhinal layer pre-α. In the transentorhinal region, such multipolar cells slope gradually downwards until they reach the upper surface of layer V (Fig. 6.2). In so doing, they undergo a transformation from star-shaped multipolar cells in layer pre-α into modified pyramidal cells with an apical dendrite that develops gradually (Braak and Braak 1992a).

These modified pyramidal cells develop tau aggregates so suddenly that it is impossible to distinguish any precursors under the light microscope. In contrast to the process in coeruleus neurons, however, abnormal tau material frequently appears initially in dendrites (Fig. 7.2b) and then fills the soma and the axon

H. Braak, K. Del Tredici, *Neuroanatomy and Pathology of Sporadic Alzheimer's Disease*, Advances in Anatomy, Embryology and Cell Biology 215,
DOI 10.1007/978-3-319-12679-1_7

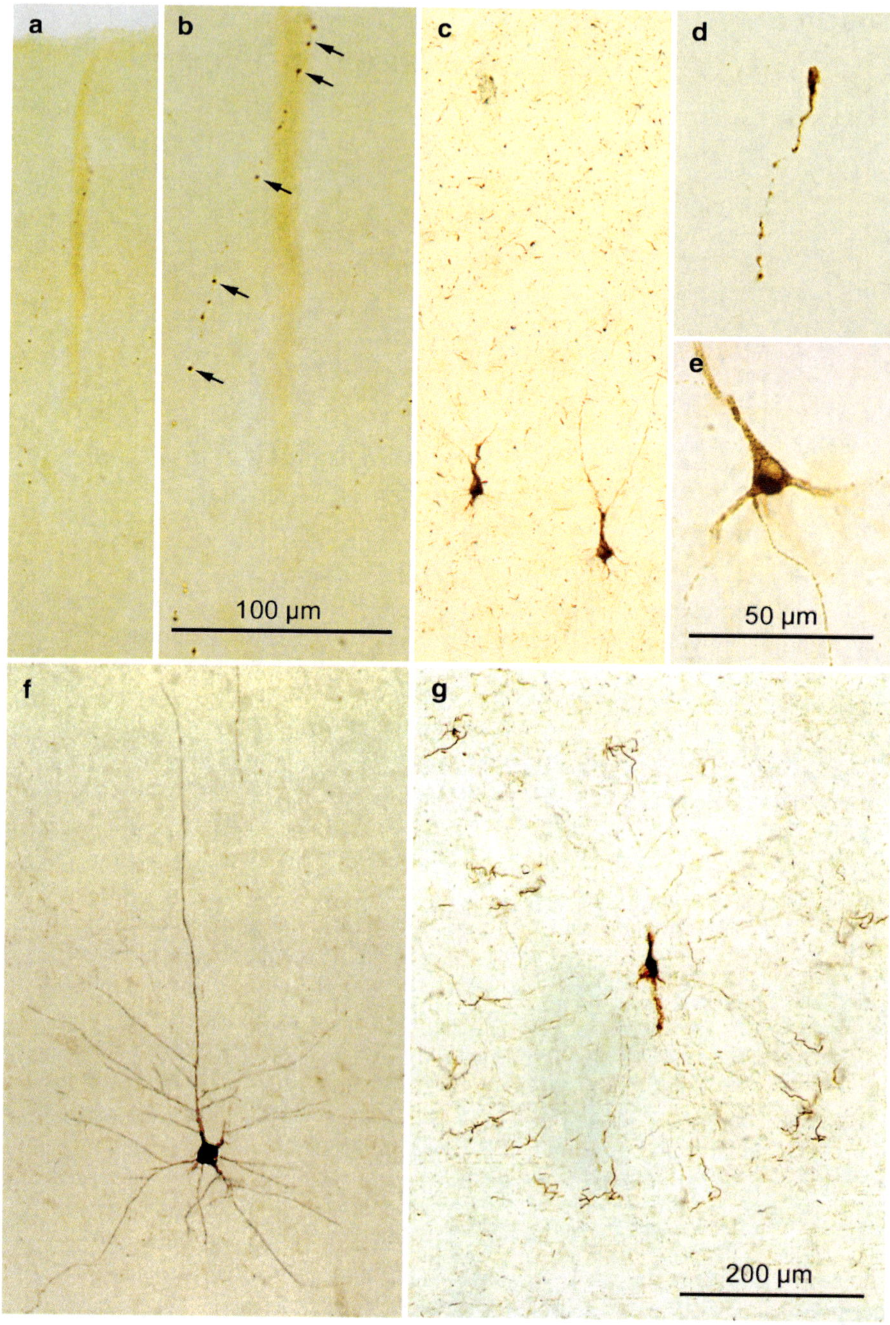

Fig. 7.1 Stages 1a and 1b (cortical AT8-ir but non-argyrophilic tau inclusions) in 100 μm AT8-immunostained sections. (**a**, **b**) Stage 1a. (**a**) At first glance it appears that the transentorhinal cortex has no AT8-ir lesions. Micrograph in (**b**) shows the region around the cortical vessel seen, in (**a**), at higher magnification. *Arrows* point to a single AT8-ir axon, with pathological inclusions

(Fig. 7.2c) (Braak et al. 1994). It is indeed difficult to comprehend how these modified pyramidal cells could synthesize enough normal tau in the somatodendritic compartment within such a short interval. As such, it is more probable that a production phase precedes the immediately beginning aggregation. Yet, it is unknown what mechanisms trigger the pyramidal cells (e.g., pathological signaling) to commence producing large amounts of normal tau because their somatodendritic compartments do not need the protein. Up to stage 1a, only the brainstem nuclei with diffuse cortical projections display AD-associated intraneuronal lesions. Thus, it is logical to surmise that transentorhinal projections neurons can be reached and influenced by the axonal terminals of these nuclei, although it is admittedly difficult to understand how such a one-to-one contact between a single axon terminal and a single neuron can take place. Since only a small fraction of the newly synthesized tau protein finds binding sites in the somatodendritic compartment, the protein presumably exists in the cytosol, for the most part, in a hyperphosphorylated state. There, after exceeding critical concentrations, it may convert within a short time interval into an irreversibly hyperphosphorylated and slightly aggregated state.

Immediately afterwards, the involved pyramidal cells are nearly filled—reminiscent of a Golgi impregnation—with the pathological material and, initially, they barely deviate from their normal shape (Figs. 7.1f and 7.2c). Next, local swellings appear at the most distal segments of the dendrites, which slowly become curled and twisted (Figs. 7.1g and 7.2c). With time, the terminal dendrites develop short appendages and then look as if they are isolated in the neuropil without any connection to their proximal portions (Figs. 7.1g and 7.2c) (Braak et al. 1994). Notably, the distal dendritic segments of cortical pyramidal cells are phylo- and ontogenetically late-appearing structures that chiefly receive axonal contacts of late-emerging and late-maturing pyramidal cells. The functions of these contacts are not known. In any case, loss of the most distal dendritic segments does not impinge on the survival of the involved neurons. At this early stage, the lost dendritic segments contain AT8-ir material that has not converted into argyrophilic inclusions (NTs). The detached dendritic segments are rapidly degraded and leave behind no remnants. At present, it is still unknown whether the abnormal material becomes a component of the interstitial fluid (ISF) and, subsequently, of the cerebrospinal fluid (CSF).

Fig. 7.1 (continued) appearing like a string of perls. (**d**) Occasionally, terminal portions of altered axons with knob-like enlargements at their tips can be observed. The immunoreactive material in axons is recognizable in thick sections but easily missed in thin paraffin sections used for routine diagnostics. (**c**, **e**–**g**) Stage 1b. AT8-ir pyramidal cells appear suddenly in addition to the already-existing subtle axonal network (see especially the upper half of **c**). (**e**) AT8-ir material is present not only in the somatodendritic domain but also in the axon of the affected pyramidal cells. (**f**) Initially, the AT8-ir pyramidal cells appear intact and resemble Golgi-impregnations of a healthy nerve cell. (**g**) Later, however, the distalmost segments of their dendrites become twisted and exhibit varicosities and appendages. Eventually, these tips lose their contact to the proximal branches. Scale bar in (**g**) also applies to (**a**, **c**) and (**f**)

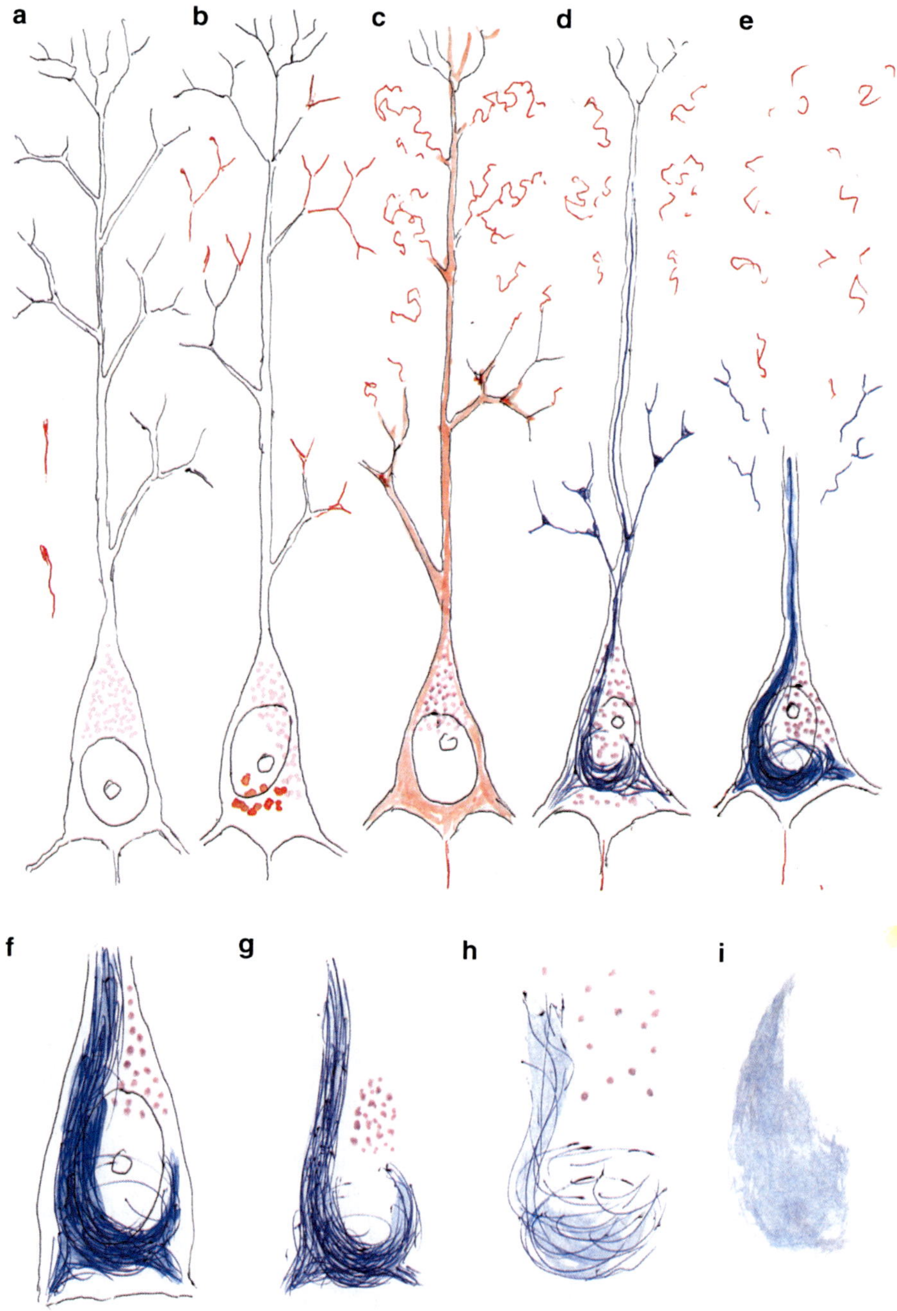

Fig. 7.2 Schematic drawing of the development of pathological changes in transentorhinal pyramidal cells of layer pre-α. (**a**) The cortical neuropil displays AT8-ir neuronal processes, some of which show a knob-like enlargement at their tip. The lipofuscin granules (*violet* stippling) usually lie close to the stem of the apical dendrite. (**b**) With relative abruptness, AT8-ir material

A similar procedure occurs in all of the different types of nerve cells that are susceptible to the AD process. That is, the somatodendritic compartments of these cells pass through the previously described pretangle phase before developing argyrophilic filaments in their somata (NFTs) and dendrites (NTs). The potential for reversing the pathological process is probably highest during the pretangle phase.

7.2 NFT Stages I and II

After passing through stage 1b of the pretangle phase, involved nerve cells begin, for the first time, to produce argyrophilic lesions at branching points of their dendrites. In a further development, the argyrophilic material extends into other portions of the dendrite and forms NTs (Togo et al. 2004). It is still fully unclear why, as a rule, the development of NTs in pyramidal cells precedes that of NFTs. A variety of flame-, club-shaped, and branching NTs develops in dendrites.

Argyrophilic particles within the cytoplasm tend, initially, to lie loosely dispersed within the deposits of lipofuscin or neuromelanin granules. The central portion of larger NFTs often forms a network around the pigment granules, and these granules may function as an initiation site for promoting further aggregation (Fig. 7.2d, e). This idea is supported by the observation that neither pretangle nor argyrophilic material are found near the Nissl substance. In addition, nerve cell types that do not produce lipofuscin or neuromelanin granules are remarkably resistant to the formation of argyrophilic NTs/NFTs (see also Sect. 9.2).

Mature NFTs are frequently flame-like or comet-like in appearance (Figs. 7.2f and 9.4e, f), whereas other NFT types are rounded or globose. Gradually, NFTs fill large portions of the soma. They even can displace the nucleus toward the periphery and they frequently protrude somewhat into the proximal stems of dendrites; however, they do not extend into the axon (Fig. 7.2d–f).

Fig. 7.2 (continued) appears intraneuronally in the form of droplets in the soma but also distributed diffusely at branching sites of dendrites. (**c**) The soluble material fills the entire cell, including the axon. Then, the distal dendritic segments become abnormally altered and detached from the stems. (**d**) The abnormal tau material in the somatodendritic compartment readily converts into argyrophilic NTs and NFTs (argyrophilic material is depicted in *blue*), whereas the abnormal material in the axon remains in a non-argyrophilic state. (**e**) The NFT consolidates and partially extends into dendritic stems. Terminal branches of the dendrites are lost, and the complexity of the dendritic tree is severely reduced. (**f**) The cell soma of surviving pyramidal cells shows a light nucleus with a large nucleolus. (**g**) Loss of the cell nucleus is seen easily in thicker sections. The compactness of the NFT gradually decreases, and the lipofuscin granules maintain their position for awhile. (**h**) After a relatively long period of time, the lipofuscin granules begin to dissipate and they eventually disappear from the tissue. The argyrophilia of the NFT decreases and the NFT loses its compact appearance. (**i**) The NFT leaves behind a permanent tombstone tangle. Here, AT8-ir material appears in red and Gallyas-positive material is depicted in *blue*

In persons of advanced age, neocortical layers IIIa and IIIb often contain pyramidal cells that have a massive spindle-shaped enlargement between basal portions of their cell body and the displaced axonal initial segment first beginning at the tip of the enlargement (meganeurites) (Purpura and Baker 1978; Braak 1979). Not only the somatodendritic compartments of involved cells but also the spindle-shaped meganeurites at these locations fill up with AT8-ir or argyrophilic material, thereby confirming the notion that argyrophilic material does not go beyond the axon initial segment into the axon.

NFT Stage I: Mild argyrophilic lesions develop in the transentorhinal region, involvement of the spinal cord and olfactory bulb

By NFT stage I, all of the non-thalamic nuclei with diffuse cortical projections have developed at least some degree of tau pathology. Along with the cholinergic projection cells of the magnocellular nuclei of the basal forebrain, one also finds AT8-ir material in nerve cells of the hypothalamic tuberomamillary nucleus. At this point, the pathological process also reaches regions of the non-thalamic nuclei that generate descending projections to the lower brainstem and spinal cord, i.e., the lower raphe nuclei and the subcoeruleus nucleus. Axonal networks containing tau aggregates also develop in regions targeted by these nuclei, including the spinal cord, where an AT8-ir axonal network and, more infrequently AT8-ir cell somata are present, decreasing in extent from the cervical to sacral segments (Fig. 9.6a) (Dugger et al. 2013).

Within the cerebral cortex, the transentorhinal region is usually the first site where NTs/NFTs in pyramidal neurons are present (Figs. 7.3a and 9.1a–c). At the same time, one always encounters projection cells there that only show pretangle material. The entorhinal region proper initially remains uninvolved or minimally involved, so that the focus of the tau pathology is clearly in the transentorhinal region. Later in stage I, abundant AT8-ir and Gallyas-positive neurons mark the descent of the superficial entorhinal cellular layer pre-α from its outermost position at the entorhinal border to its deepest position at the transition towards the adjoining temporal neocortex (Fig. 7.3b).

Tau pathology occurs at this stage in the anterior olfactory nucleus as well as in tufted and mitral cells of the olfactory bulb, and it increases with disease progression (Fig. 9.6d–f). In subsequent stages, the superordinate cortical components of the olfactory system, including the olfactory portions of the amygdala (cortical subnuclei) and of the entorhinal region (ambient gyrus) become involved. Inasmuch as Aβ deposition in the olfactory bulb is first present in stage III (Attems and Jellinger 2006), the primary pathology there during AD is tangle formation (Kovács et al. 1999).

NFT Stage II: Argyrophilic lesions progress into the entorhinal region

From the transentorhinal cortex, Gallyas-positive lesions encroach upon the entorhinal region and particularly the superficial cellular layer pre-α (Figs. 7.3c and 9.1d–f). Subsequently, silver-stained lesions also develop in the deep pri-α layer. Both layers gradually become macroscopically visible in thick sections (Fig. 7.3c). Layer pri-α shows sharply defined upper and lower boundaries and it

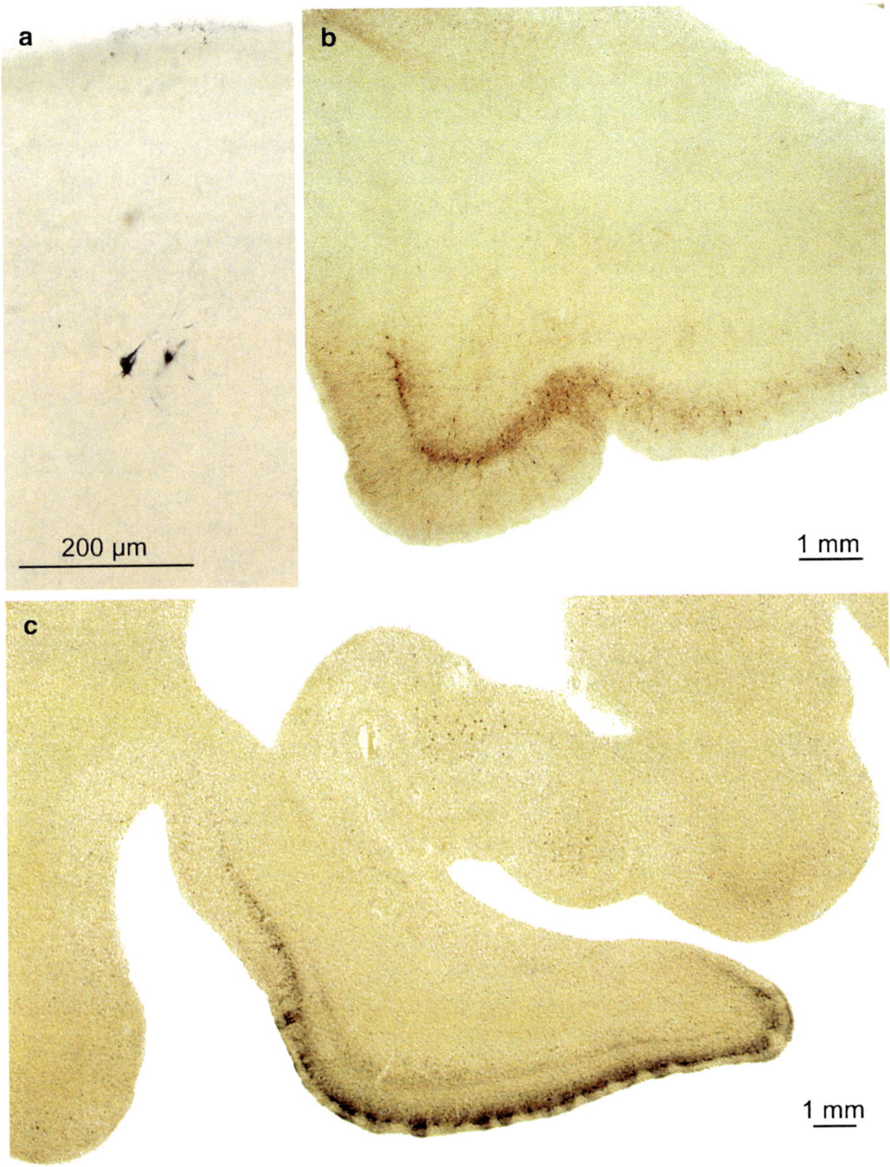

Fig. 7.3 NFT stages I and II in 100 μm sections. (**a**) Stage I: The transentorhinal region with initial cortical pathology, here in two Gallyas-positive cells. Such subtle lesions can easily be overlooked (49-year-old female, Gallyas silver-iodide impregnation). (**b**) Advanced NFT stage I: Very large numbers of altered layer pre-α pyramidal cells are seen chiefly in the transentorhinal region, and only a minority of them extend into the entorhinal region proper (AT8 immunoreaction). (**c**) NFT stage II: Here, not only the transentorhinal region (note the oblique course of the pre-α layer) but also the entire entorhinal region up to the ambient gyrus has become involved. This section also shows the beginning involvement of the deep layer pri-α and subtle involvement of uncal portions of the hippocampal formation (Gallyas silver-iodide impregnation)

is separated from external layers and sometimes also from the following deep layer, pri-β, by the broad, wedge-shaped *lamina dissecans*, a myelinated fiber plexus (Fig. 7.3c) (Braak and Braak 1992a; Insausti and Amaral 2012). The presubicular region is uninvolved at this stage. In the hippocampal formation, AT8-ir pyramidal cells begin to appear either in CA 1 or in CA 2. Varicosities filled with abnormal tau material are transiently observable in apical dendrites that pass through the *stratum lacunosum moleculare* of CA 1 (Braak and Braak 1997a) and lie directly along the way of the perforant pathway (Fig. 9.5a, b). Filigrane networks of AT8-ir axons and dendrites develop in the *stratum radiatum* and *stratum oriens* (Fig. 9.5a, b).

7.3 Prevalence of Stages a–II

The relationships among age and AD-associated tau pathology can be studied by staging relatively large numbers of non-selected autopsy cases (Braak et al. 2011). The columns in Fig. 7.4 show the prevalence in n = 1,895/2,366 non-selected cases ranging in age from 1 to 100 years and ranging from those with no tau pathology to those at stages a–II (Table 7.1; Fig. 7.4). The columns represent decades, and the number of cases in each decade that reached the stage(s) shown in each of the four graphs is indicated above the columns (Fig. 7.4a–d). The prevalence of the lesions depicted in Fig. 7.4 does not represent standard epidemiological data because the data were not collected from a living population. Their interpretation rests upon the assumption that the pathological process in AD progresses from stage a to stage VI as a continuum.

Remarkably, only a few cases show the complete absence of abnormal tau inclusions in the CNS (10/2,366 cases = less than 1 % of all cases) (Table 7.1; Fig. 7.4a). Six of these ten cases are younger than 10 years of age and all of them are devoid of Aβ deposition (Table 7.1). The inclusion of 'negative' results here and below is not unproblematic: In contrast to a positive finding, the validity of a negative finding is subject to limitations, e.g., here, assessment of a single 100 μm tissue section for each region. Where a negative finding results, it cannot be ruled out that, under other circumstances or conditions (control of more sections), the outcome may have been different.

As seen in Fig. 7.4b–d, more than 99 % (and, as of the fourth decade, all) of the individuals sampled show the presence of early tau pathology in one or the other stage. These intraneuronal lesions in the human CNS are not benign or 'normal' because they become increasingly severe and cause cellular dysfunction ultimately leading to the premature death of the involved neurons.

Cases in Fig. 7.4b exhibit the presence of AT8-ir material only in the locus coeruleus and/or other nuclei with ascending and diffuse cortical projections (subcortical stages a–c: 57 of all 2,366 cases = approximately 2 %) (Table 7.1; Fig. 7.4b). The cerebral cortex and other brain regions are devoid of AT8-ir material. Cases characterized by the three lesional distribution patterns 'a' (tau aggregates in axons of the locus coeruleus), 'b' (tau aggregates in melanized

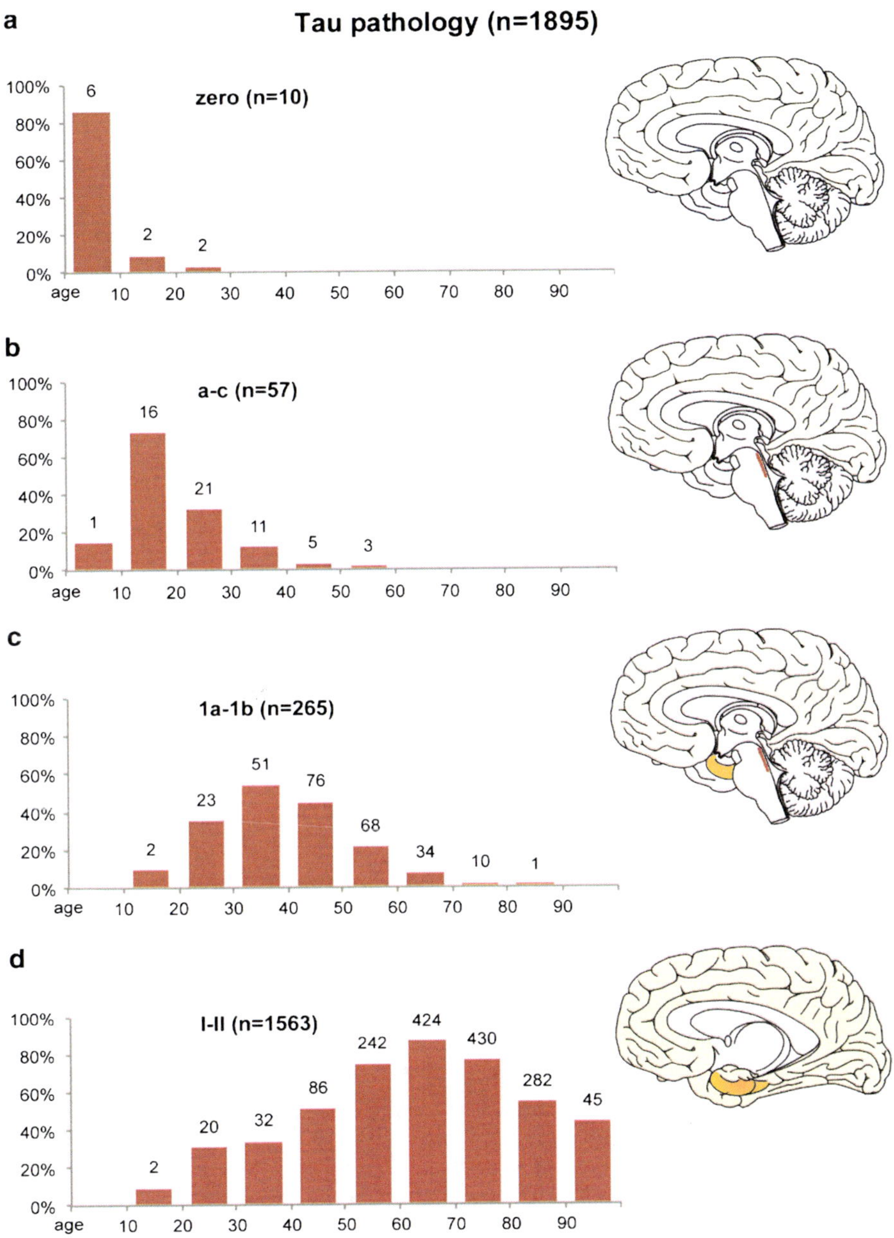

Fig. 7.4 Development of early abnormal intraneuronal tau deposits in n = 1,895 of n = 2,366 non-selected autopsy cases according to decades (ages of the cohort 1–100). Columns display the frequency of cases in relation to the total number of cases in the various age categories. (**a**) The first row displays the prevalence of cases lacking AD-associated tau aggregates. (**b**–**d**) These rows show the evolution of the intraneuronal tau changes: non-fibrillar AT8-ir and still non-argyrophilic aggregates in subcortical stages a–c (**b**), as opposed to AT8-ir, still non-argyrophilic, aggregates in the cerebral cortex in stages 1a and 1b (**c**). (**d**) Argyrophilic neurofibrillary lesions in cortical nerve cells are characteristic of NFT stages I and II

Table 7.1 Development of early tau pathology in n = 1,885 of a total of n = 2,366 cases according to decades, including ratio between females (n = 774) and males (n = 1,111)

Age (n)	Zero (AT8)	a–c (AT8)	1a–1b (AT8)	NFT I (Gallyas)	NFT II (Gallyas)
0–9 n = 7	6 (2/4) 85.71 %	1 (0/1) 14.28 %	0 0 %	0 0 %	0 0 %
10–19 n = 22	2 (0/2) 9.09 %	16 (3/13) 72.72 %	2 (0/2) 9.09 %	2 (0/2) 9.09 %	0 0 %
20–29 n = 66	2 (2/0) 3.03 %	21 (9/12) 31.81 %	23 (9/14) 34.84 %	18 (8/10) 27.27 %	2 (2/0) 3.03 %
30–39 n = 95	0 0 %	11 (3/8) 11.58 %	51 (28/23) 53.68 %	28 (12/16) 29.47 %	4 (0/4) 4.21 %
40–49 n = 170	0 0 %	5 (3/2) 2.94 %	76 (28/48) 44.70 %	81 (35/46) 47.65 %	5 (3/2) 2.94 %
50–59 n = 326	0 0 %	3 (0/3) 0.92 %	68 (28/40) 20.86 %	219 (72/147) 67.18 %	23 (1/22) 7.06 %
60–69 n = 487	0 0 %	0 0 %	34 (9/25) 6.98 %	305 (107/198) 62.63 %	119 (42/77) 24.44 %
70–79 n = 564	0 0 %	0 0 %	10 (4/6) 1.77 %	222 (93/129) 39.36 %	208 (100/108) 36.88 %
80–89 n = 525	0 0 %	0 0 %	1 (0/1) 0.19 %	94 (48/46) 17.90 %	188 (101/87) 35.81 %
90–100 n = 104	0 0 %	0 0 %	0 0 %	11 (6/5) 10.58 %	34 (20/14) 32.69 %
Total n = 2366	10 (4/6) 0.42 %	57 (18/39) 2.41 %	265 (106/159) 11.20 %	980 (381/599) 41.42 %	583 (269/314) 24.64 %

neurons of the locus coeruleus), and 'c' (tau aggregates in the locus coeruleus and in other subcortical nuclei with diffuse cortical projections) occur, surprisingly, at very young ages. The fact that irreversibly hyperphosphorylated and slightly aggregated tau in axons and pretangle material in the somatodendritic compartment develop in brainstem nuclei of young individuals, implies that advanced age alone is not mandatory for the production of the pathological material. Overview sections stained for lipofuscin pigment and basophilic material do not reveal any obvious pathological alterations (e.g., loss of basophilic material, displacement of cell nuclei to the periphery). The prevalence of such cases culminates in the second decade and slowly decreases thereafter, although these stages are present until the end of the sixth decade (Table 7.1; Fig. 7.3b). The intraneuronal pathology during the early stages a–c is not accompanied by insoluble extracellular deposits of Aβ (Braak et al. 2011; Braak and Del Tredici 2011, 2012).

The brainstem nuclei with diffuse cortical projections differ essentially from other systems within the CNS. As such, it should become possible in the future to detect even incipient or slight disturbances within the noradrenergic and other systems by developing new and sensitive non-invasive tests. Systemic abnormalities potentially caused by very early AD-associated lesions in young persons (for instance, early dysregulation of the intracerebral microvascular system and the blood brain barrier) may have been overlooked or misinterpreted up to now

Table 7.2 Comparison of early stages of AD-associated tau lesions with phases of Aβ deposition (n = 1,885)

Aβ phases	Tau zero	Tau stages a–c	Tau stages 1a–1b	NFT stage I	NFT stage II	Stages a–II
Zero	10	57	244	687	265	1,253
1	0	0	12	141	103	256
2	0	0	9	128	135	272
3	0	0	0	22	74	96
4	0	0	0	2	6	8
5	0	0	0	0	0	0
Total	10	57	265	980	583	1,885

because, from a differential diagnostic standpoint, they may not have been viewed within the context of the AD-process.

A third graph shows a total of 265 cases in which, in addition to subcortical lesions, cortical tau pathology occurs for the first time, above all in the transentorhinal region (stages 1a and 1b) (Table 7.1; Fig. 7.4c). This intraneuronal material is AT8-ir but non-argyrophilic. In stage 1a (38/2,366 cases = approximately 2 %), it is confined solely to neuritic processes, very probably extrinsic axons. Also included in this graphic are cases with cortical AT8-ir pyramidal cells that mostly represent modified layer pre-α cells of the transentorhinal region (stage 1b: 227/2,366 cases = approximately 9 %) (Table 7.1; Fig. 7.4c). Some individuals display only a single involved pyramidal cell, whereas others exhibit more AT8-ir neurons. All 1a and 1b cases (265/2,366 cases = approximately 11 %) have patterns of subcortical lesions resembling those in stages a–c; in other words, the cortical tau lesions do not occur in the absence of subcortical tau pathology. The prevalence of cases in stages 1a and 1b increases steadily up to the fourth decade. Thereafter, it slowly decreases up to the ninth decade (Table 7.1; Fig. 7.4c). Some individuals at stages 1a/1b also develop initial Aβ deposits (12 at phase 1 = approximately 4 % of the 265 1a/1b cases and 9 at phase 2 = approximately 3 % of the 265 1a/1b cases) (Table 7.2). See also Chap. 8.

The formation of argyrophilic (Gallyas-positive) cortical lesions characterizes the following NFT stages I–VI. Brainstem nuclei also develop argyrophilic lesions, but they are not currently included into the criteria for diagnosing stages I–VI (Braak and Braak 1991a; Montine et al. 2012). During stage I, argyrophilic neurofibrillary pathology is, once again, predominantly found in the transentorhinal region (stage I: 980/2,366 cases = approximately 41 % of all cases) (Table 7.1; Fig. 7.4d). In some instances, the pathology is confined to no more than an isolated NFT-bearing pyramidal cell in layer pre-α (Fig. 7.3a) (Braak et al. 2011). The number of involved neurons there increases, and the pathological process gradually extends into layer pre-α of the entorhinal region; however, in stage I, the transentorhinal region is typically the focus of the pathology (Fig. 7.3b). By contrast, in stage II, the involved nerve cells are distributed throughout layer pre-α of the transentorhinal and entorhinal regions, although the lesions within the entorhinal region gradually gain a detectable and distinct priority (stage II: 583/2,366 = approximately 25 %) (Table 7.1; Fig. 7.4d). Occasionally in stage I/II cases, a few pyramidal cells in CA 1 are also involved. The prevalence of NFT stage I and II cases combined (1,563/2,366 = approximately 66 %) rises steadily up to the seventh decade (Table 7.1; Fig. 7.4d). After that, NFT stage I/II cases decrease in frequency—only to become replaced by higher NFT stages (Fig. 9.13a).

NFT stage I is often accompanied by the extracellular Aβ deposition, mostly phase 1, but also up to and including phase 4 (293/980 cases = approximately 30 %) (Table 7.2). As expected, stage II cases display Aβ deposition more frequently than cases at stage I, with most reaching phase 2; but some reach phase 4 (317 cases = approximately 54 %) (Table 7.2). Individuals with tau lesions corresponding to stages a–II as well as those with stages I–II plus Aβ deposition do not manifest AD-related symptoms or they fall below the clinical detection threshold using currently available diagnostic means.

Owing to the nature of the pretangle lesions, stages a–c and 1a–1b can only be assessed using AT8 immunoreactions. However, as soon as *argyrophilic* tau inclusions begin to appear, this situation changes: It is the argyrophilic tau aggregates that count for staging purposes and the stages are designated by the Roman numerals I-VI. The assessment of stages in Gallyas silver-stained sections differs from that in AT8-immunoreactions performed on the same cases (Table 7.3). In many cases, both staging scores are identical, e.g., scores of Gallyas-stained (G) sections show the same staging result as scores of AT8 immunoreactions (A): 612/980 stage I cases = approximately 62 %, and 207/583 stage II cases = approximately 36 % (Table 7.3). For other cases, however, both staining methods result in a variance of one stage: 306/980 stage I cases = approximately 31 %, and 348/583 stage II cases = approximately 60 %). In a small number of cases, the variance amounts to two stages: 62/980 stage I cases = approximately 6 %, and 28/583 stage II cases = approximately 5 % (Table 7.3). These differences need to be taken into consideration when only immunohistochemistry is performed for routine purposes to minimize time.

Assuming the course of the pathological process proceeds from beginning to end at nearly the same rate, one would anticipate that the argyrophilic lesions constantly evolve—after approximately the same amount of time has elapsed—out of the non-argyrophilic lesions. If this were to be true, the discrepancy between both staging scores should not exceed one NFT stage, and it should be possible, using two endpoints (i.e., the presumed beginning of the tau lesions during, say, the first or second decade of life and the tau stage at the time of death) to estimate the approximate time point at which the AD process would have reached a clinically recognizable threshold, e.g., NFT stage V (Ohm et al. 1995). In fact, however, a small group of cases displays a discrepancy of two NFT stages. In such individuals, it may be that, during the last phase of their illness prior to death, the tempo of the pathological process accelerates, with the result that the non-argyrophilic lesions henceforth outnumber the argyrophilic pathology and, as such, dominate the overall

Table 7.3 Differences between Gallyas (G) and AT8 (A) scores for NFT stage I cases (left) and NFT stage II cases (right)

Age	n = G I	G = A	= I + 1	= I + 2	I + 2	= %	n = G II	G = A	= II + 1	= II + 2	I + 2	= %
10–19	2	2	0	0	0	0	0	0	0	0	0	0
20–29	18	16	2	0	2	11	2	1	1	0	1	50
30–39	28	27	1	0	1	4	4	2	1	1	2	50
40–49	81	71	10	0	10	12	5	3	1	1	2	40
50–59	219	168	46	5	51	23	23	9	13	1	14	61
60–69	305	183	105	17	122	40	119	64	50	5	55	46
70–79	222	113	84	25	109	49	208	74	126	8	134	64
80–89	94	27	54	13	67	71	188	47	131	10	141	75
90–100	11	5	4	2	6	55	34	7	25	2	27	79
Total	980	612	306	62	368	38	583	207	348	28	376	64

picture. Currently, no one knows which mechanisms cause or influence these phenomena. Nevertheless, Table 7.3 shows that the proportion of cases with discrepant AT8 and Gallyas scores increases in the later decades of life and at higher tau stages.

The existence of many elderly subjects with stages I–II shows that it is not only possible to reach old age with mild lesions but also that the rates at which the AD process progresses differ remarkably from one individual to another (Fig. 7.4d). Some reach stages I–II as teenagers or in early adulthood; others have to be over 90 years of age to do so. This means that the pathological process underlying AD does not inevitably lead to dementia (Ferrer 2012). Rather, as a rule, it fails to attain the dimensions that would lead to clinically recognizable symptoms. Currently, these marked inter-individual differences cannot be explained adequately. However, more information regarding all of the factors that determine and influence the tempo of the pathological process is urgently needed.

Some of the data in this monograph perhaps convey the impression that, with increasing age and given the ubiquitous occurrence of the AD process, a curtailment of higher CNS functions is unavoidable. For this reason, it should be emphasized that a life-long maturation process on the part of the CNS militates against such a development. Although this maturation process is not the subject of the present work, nevertheless, it contributes to enduring improvements in the functioning and potential of the CNS. Both of these counter-trends together with their different time frames are subject to numerous additional factors and cause human brain structure and function to undergo change continually over time (Braak et al. 2003; Johnson et al. 2009). The potential of this remarkable organ remains individually distinctive and unique.

7.4 The Problem of Selective Vulnerability and the Potential Transmission of Pathological Changes from One Neuron to the Next

One of the remarkable aspects about disease progression in AD is that brain sites become involved in a predictable sequence with relatively little inter-individual variability (Kemper 1984). At present, there is no patent explanation for the sequential and regional progression of the pathological process. One option is to assume a differential and preferential vulnerability of all of the diverse neuronal types involved, i.e., the pathogenesis of tau aggregates in a cell-autonomous manner; however, the mechanisms underlying this presumed differential vulnerability have yet to be identified. The extreme differences in the degrees of vulnerability among varieties of one and the same basic type of nerve cell are particularly problematic, as is the case, for instance, with cortical pyramidal cells: Those in layer Va are particularly prone to develop the lesions, whereas those in the suprajacent layer IV are resistant.

Often overlooked is that the hypothetical construct of differential vulnerabilities in AD fails to take into account that all involved brain regions and all involved neuronal types are anatomically interconnected. This interconnectivity indicates that physical contacts between involved regions play a key role in the pathogenesis of AD (Saper et al. 1987; Pearson and Powell 1989; Pearson 1996). In fact, routes exist that would permit a continual propagation of AD pathology via anterograde axonal transport and transsynaptic transmission of tau by means of an exchange of appropriate pathological signals (Dujardin et al. 2014), and work is being done on anterograde transmission of endogenous pathogenic molecules or on experimental models for an exchange of altered signals between one involved neuron and the next interconnected but still uninvolved neuron (Trojanowski and Lee 2000; Clavaguera et al. 2009, 2013a, b, 2014; Guo and Lee 2011, 2014; de Calignon et al. 2012; Liu et al. 2012; Duyckaerts 2013; Walker et al. 2013; Ahmed et al. 2014; Medina and Ávila 2014a).

The lower brainstem predilection site (locus coeruleus) is located at a considerable distance from the cortical transentorhinal region, the next site where AD-associated tau pathology develops, and axons originating from the coeruleus project to this region (Fig. 6.10). As such, only the terminal segment of long axons and possibly only synapses with both pre- and postsynaptic sites are candidates for potential propagation (Liu et al. 2012). Although no data are presently available as to how non-junctional varicosities operate, one can see that the pathological process initially is confined to cortical pyramidal cells in the transentorhinal region.

It is conceivable that defective axons send aberrant signals to the cortical nerve cells on which they synapse and that these signals induce pathological changes within the tau protein. Alternatively, small doses of pathogenic molecules (for instance, soluble but irreversibly hyperphosphorylated and slightly aggregated tau molecules) could be released into the synaptic cleft and taken up at the postsynaptic site by the recipient neuron. Once inside the receptor cell, the tau molecules could act as seeds triggering the production of abnormal tau there. The conditions needed to promote or enhance such signaling or seeding are incompletely understood, but there is increasing experimental evidence for the working hypothesis that aggregated tau can be transferred from one nerve cell to another and that such aggregates can induce the production of new abnormal tau in receptor cells (Goedert et al. 2010, 2014; Lee et al. 2010; Jucker and Walker 2011; Iba et al. 2013; Kaufman and Diamond 2013; Van Ba et al. 2013; Holmes et al. 2014; Medina and Ávila 2014b; Sanders et al. 2014).

Phylogenetic influences may be partially responsible for the spread of the pathological process to neuronal constituents of the cerebral cortex just within the reaches of the transentorhinal region. As pointed out previously, the transentorhinal region functions as an interface between the phylogenetically and ontogenetically late-developing basal temporal neocortex and neocortically-oriented portions of the entorhinal region. In primates, the transentorhinal region increases in size and its topographical extent peaks in humans (Braak and Braak 1992a). These phylogenetically late events are reflected by similar developments in the subcortical nuclei that project to these cortical regions (integrated phylogeny) (Rapoport 1988, 1989, 1990, 1999). Therefore, the most recent 'acquisitions' of the locus coeruleus probably are nerve cells that project to the transentorhinal region. These circumstances may help to explain why these two neuronal types (coeruleus melanized neurons and transentorhinal pyramidal cells) normally develop the first AD-associated tau lesions within the entire CNS (Braak and Del Tredici 2011, 2012).

The distal axonal segment of projection neurons in the locus coeruleus is heavily ramified. If the same morphology applies for all of these axons' side branches, then precisely localized one-to-one effects are inconceivable. Broadly ranging effects are chiefly produced by means of non-junctional varicosities (e.g., the release of noradrenalin into the ISF). Complete synapses, by contrast, produce local effects. It is possible that the primary branches of individual coeruleus axons differ from their side branches with respect to dissimilar supplies of synaptic terminals. However, the question of how heavily ramified axons produce locally limited effects is a completely open one.

It may well be that it is not the insoluble Gallyas-positive fibrillar material in the somatodendritic compartment of involved neurons that is harmful. Likewise, it is unlikely that dangerous material could be released from dendrites or the somata of coeruleus neurons because the next nerve cell in which the lesions develop would have to be in the immediate vicinity of affected cells in the locus coeruleus. Yet, this is not the case. Instead, it appears that only newly involved axons possess for a limited time some reserves of soluble and slightly viscous (non-argyrophilic) abnormal tau that could be the real candidate for the potential spread of the AD process.

The frequently seen early involvement of the second sector of the Ammon's horn may also be attributable to axons projecting from subcortical nuclei with diffuse cortical projections. The terminals from the perforant pathway that synapse on CA 2 are meager, thereby making the former an improbable route of tau propagation. By contrast, CA 2 is the recipient of projections from the hypothalamic tuberomamillary nucleus. Involvement of this nucleus could induce the AD-associated lesions in CA 2, which also typically display a certain temporal independence from the perforant pathway-mediated tau lesions in CA 1.

The hypothesis of a neuron-to-neuron seeding and propagation via synapses with pre- and postsynaptic sites offers a straightforward explanation both for the predictable regional distribution pattern of the tau lesions and the slow rate of disease progression (Braak et al. 2011). Knowledge of the underlying mechanisms would mean that they could be influenced by slowing or interrupting the spread of the pathological process. The prospect of developing a causal therapy for AD during the phase when the process is confined to the brainstem and prior to the involvement of the cerebral cortex is challenging but certainly worthwhile.

Neuron-to-neuron spreading has also been discussed apropos other neurodegenerative disorders (Frost and Diamond 2010; Goedert et al. 2010, 2014; Jucker and Walker 2011; Prusiner 2012). As in sporadic AD, the pathological process in sporadic PD also progresses in a systematic manner from one region to another (Braak and Del Tredici 2009). A major difference, however, between the two diseases is that the initial α-synuclein aggregates develop elsewhere within the CNS. In PD, the olfactory bulb and the dorsal motor nucleus of the vagal nerve become involved early. The efferences of the dorsal motor nucleus connect it closely with the ENS and PNS, whereas the projections of the locus coeruleus only reach sites within the CNS. These important features may account for the noticeable differences between the regional distribution patterns and the progression of both intraneuronal pathologies. Whereas in PD the lesions can be found in all portions of the nervous system (CNS, ENS and PNS), they remain mostly confined to the CNS in AD.

7.5 Imaging Techniques and Soluble Tau as Biomarker in the CSF

Neuroimaging techniques for the clinical detection of pathological tau in the brain are in the earliest phase of development (Fodero-Tavoletti et al. 2011, 2014; Jensen et al. 2011; Maruyama et al. 2013; Okamura et al. 2013; Perani 2014; Tago et al. 2014; Villemagne et al. 2014). Suitable biomarkers, including those detectable in the CSF, are constantly being sought to investigate the presence of AD pathology in patients with the goal of monitoring disease progression. The fluid is continuously produced by all cells of the CNS. It is given off by them into the ISF filling the extracellular space, and is continuously drained into the CSF and the venous and lymphatic systems. In addition, CSF is also produced in the choroid plexus and given off from there into the ventricles. Low concentrations of soluble tau are always found in the CSF and, at higher concentrations, in the ISF.

Tau is an intraneuronally produced protein and it is normally tightly bound to the microtubules of axons. The mechanisms by which normal tau enters via the ISF into the CSF are not known (Hall and Saman 2012; Lee et al. 2012a). Along with monomeric normal tau, small amounts of monomeric hyperphosphorylated tau are found in the CSF and, generally, they are also present in the axoplasm of healthy neurons. When the production of irreversibly hyperphosphorylated, slightly aggregated AT8-ir tau begins in axons, the level of hyperphosphorylated tau protein in the CSF increases. The sources of this material are unclear, although one conceivable scenario could be that hyperphosphorylated tau is released into the ISF/CSF via volume transmission from presynaptic varicosities of axons already involved in the AD process. Another source could be the distal dendritic segments of already involved pyramidal cells. Conceptually more difficult would be a discharge of hyperphosphorylated tau from the heavily aggregated and partially truncated material belonging to dead nerve cells. Tombstone tangles remain stable for decades after the host cells die and are subject to only very gradual changes under the influence of macrophages (Braak and Braak 1991a). Currently, no evidence exists to support a contribution by tombstone tangles to levels of soluble tau in the CSF (Braak et al. 2013, but see Hall and Saman 2012).

The existence of tau as a component of the ISF/CSF—regardless in what state of phosphorylation—requires either disrupted dendritic processes, damaged axonal membranes, or the existence of specific release mechanisms (Hall and Saman 2012; Lee et al. 2012a). The total amount of soluble tau or 'T-tau' is an inter-individually comparable level of variably phosphorylated tau in the CSF that can be monitored over longer periods of time (Blennow et al. 2007; Zetterberg et al. 2007; Buchhave et al. 2012). In the course of the AD process, the T-tau level increases to around three times of that seen in the cognitively normal elderly (Blennow and Hampel 2003). Increases in CSF T-tau levels generally are thought to reflect increases in the severity or intensity of destructive processes in CNS nerve cells (Buerger et al. 2006; Blennow et al. 2012; Vanderstichele et al. 2013).

In early phases of the AD process, a quantifiable elevation of T-tau is detectable in the CSF (Andreasen et al. 1999; Craig-Schapiro et al. 2009; Mattson et al. 2009, 2012; Hampel et al. 2010). Steep increases of T-tau are associated with rapid progression from mild cognitive impairment to dementia (Blom et al. 2009) as well as with a high mortality rate in clinical AD (Sämgård et al. 2010). In addition, CSF T-tau correlates with the amount of post-mortally evaluated neurofibrillary pathology (Tapiola et al. 2009). However, autopsy-controlled prospective studies are needed to confirm the reliability of currently used biomarkers (Jack 2012; Jack and Holtzman 2013).

Chapter 8
Alzheimer-Associated Pathology in the Extracellular Space

8.1 The Amyloid Precursor Protein and the Abnormal Protein Aβ

A clear indicator for the end of the unusually protracted initial phase of the AD-associated pathological process is the abrupt appearance of an additional protein that appears in soluble form in the ISF: the small, i.e., 38–43, but mostly 40 or 42, amino acid-containing hydrophobic amyloid-β (Aβ) protein that at first is diffusely distributed in a monomeric state in a few circumscribed regions of the ISF but then rapidly forms insoluble aggregations, most of which are plaque-like entities. These Aβ-plaques develop with such consistency in the course of AD that they constitute one of its hallmark lesions (Masters and Selkoe 2012).

The pathological Aβ peptide is generated by abnormal proteolytic processing of a physiological constituent of the nerve cell membrane, the amyloid precursor protein (APP) (Beyreuther and Masters 1991; Mattson 2004; Rajendran and Annaert 2012). APP is an integral membrane glycoprotein that presumably functions as a receptor. In addition, APP has been ascribed neurotropic and neuroprotective properties (Selkoe 1994; Selkoe et al. 2012).

For the most part, APP is degraded without a trace by a process that does not permit Aβ production (Fig. 8.1a). During this process, α-secretase splices the APP and generates a soluble molecule (APPsα) that is released into the ISF. The remaining membrane-bound fragment (C83) is spliced by γ-secretase, and an additional non aggregation-prone fragment (P3) is released into the ISF, whereas the leftover APP C-terminal domain (AICD) remains in the neuronal cytoplasm (Fig. 8.1a) (Haass et al. 2012).

Aβ comes into existence only under pathological conditions and originates via an abnormal degradation pathway. First, a long and soluble fragment (APPsβ) is cleaved from APP by a β-secretase (Fig. 8.1b). The membrane-anchored fragment (C99) is subject to further clearance via γ-secretase, and the result is the release of

H. Braak, K. Del Tredici, *Neuroanatomy and Pathology of Sporadic Alzheimer's Disease*, Advances in Anatomy, Embryology and Cell Biology 215,
DOI 10.1007/978-3-319-12679-1_8

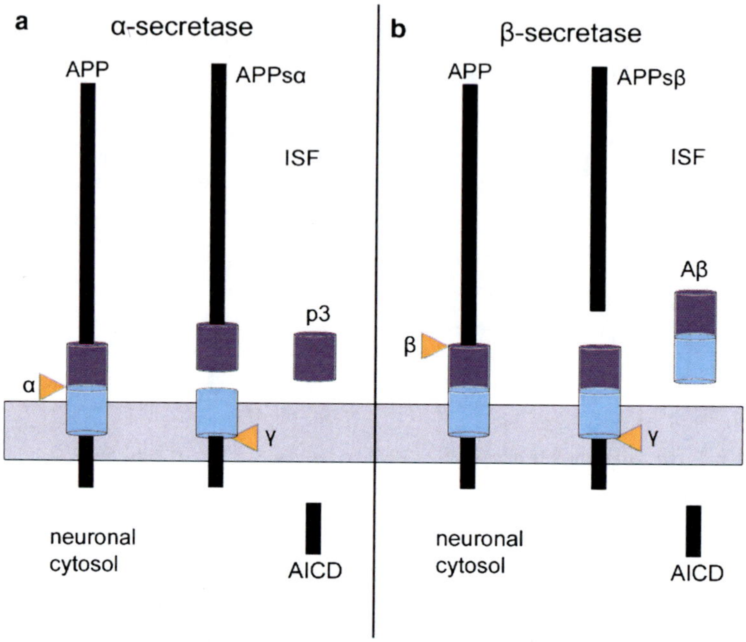

Fig. 8.1 Two processing pathways for the amyloid precursor protein. (**a**) The normal pathway utilizing α-secretase prevents the formation of Aβ and only produces p3 while, in (**b**), processing with β-secretase leads to the production of Aβ. The pathway displayed in (**b**) only occurs in a few vulnerable types of nerve cells. Diagrams adapted and reproduced with permission from C Haass et al., Trafficking and proteolytic processing of APP. Cold Spring Harb Perspect Med 2012; 2: a006270. Abbreviations: *AICD* APP intracellular C-terminal domain, *APP* amyloid precursor protein, *APPsα* soluble α-remnant of APP, *APPsβ* soluble β-remnant of APP, *ISF* interstitial fluid

Aβ into the ISF, whereas the leftover AICD remains in the neuroplasm. This sequential cleavage by β- and γ-secretases is thought to occur in the weakly acidic environment of recycling endosomes (Haass et al. 2012).

Because these steps all take place within nerve cells, the interpretation of experimental results emerging chiefly from non-polarized cells is problematic. Nonetheless, polarized cell models show that the enzymes α- und β-secretase can be distributed very differently, so that it is plausible that Aβ production by means of β-secretase can occur only at specific and predetermined sites and only in select types of nerve cells. By contrast, as anticipated, the degradation process via γ-secretase takes place at all APP-cleavage sites (Haass et al. 2012). Moreover, it is known that APP undergoes vesicular anterograde transport within axons. Thus, terminal axons and preferably presynaptic varicosities could turn out to represent the major secretion sites of Aβ (Lazarov et al. 2005).

8.2 Sources and Secretion of Aβ

Previous findings have shown that the Aβ peptide does not enter the ISF from the serum, from the vasculature, ependymal organs, or the choroid plexus. In addition, neither astrocytes, oligodendrocytes, nor microglia cells generate Aβ (Beyreuther and Masters 1991; Fiala 2007). The current consensus is that nerve cells are the sole sources of Aβ. It is very unlikely, however, that essentially *all* types of nerve cells within the nervous system produce Aβ because Aβ plaques are found only in portions of the CNS and not in the ENS or PNS. In addition, Aβ deposits do not occur with the same frequency or severity in all regions of the CNS (see also Sect. 8.5). Thus, similar to tau aggregation, Aβ deposition occurs in the CNS only at specific sites and according to a consistent developmental distribution pattern (Braak and Braak 1991a; Thal et al. 2002).

Generally, Aβ deposits in AD rarely develop in the white substance; instead, they mainly occur in the gray matter, including nerve cell somata and cellular processes of nerve cells (Figs. 8.2–8.4). In the gray matter, it is possible to distinguish regions with high densities of Aβ plaques, e.g., the anterior olfactory nucleus and olfactory bulb (Kovács et al. 1999; Attems and Jellinger 2006), the entire neo- and allocortex (Thal et al. 2002), claustrum, striatum (Braak and Braak 1990; Beach et al. 2012b), thalamus, mesencephalic tectum, red nucleus, cerebellar cortex (Braak et al. 1989b), and specific locations of the lower brainstem, from sites where Aβ plaques are sparse, e.g., the multiform layer of the neocortex, the lateral geniculate body of the thalamus, substantia nigra, and the precerebellar nuclei in the brainstem, among others. Aβ plaques are absent in both segments of the pallidum as well as in the hypothalamic lateral tuberal and lateral mamillary nuclei. This pattern of Aβ plaques occurs with little inter-individual variability and is the major reason for surmising that not all types of nerve cells of the CNS can produce Aβ.

As pointed out earlier (Sect. 2.2), CNS neurons can have a long or a short axon (Fig. 2.1e–g). The characteristic Aβ distribution pattern associated with the AD process makes it improbable that nerve cells with a short axon contribute to Aβ production because, were this to be true, one should see precipitations of Aβ in the immediate vicinity of these cells; but that does not happen. Therefore, the number of CNS nerve cell populations that produce Aβ cannot be, by process of elimination, very large. Of course, the question arises whether all projection neurons with a long axon can generate Aβ under normal conditions. If so, an ongoing Aβ production should be detectable throughout the lifetimes of all individuals irrespective of their cognitive status. However, inasmuch as there is no evidence for such a generalized process, it is clear that Aβ production is integral to the AD-associated process. Presumably, the homeostasis of projection neurons that have AD-associated intraneuronal lesions is not unperturbed. For this reason, it is possible that Aβ originates chiefly, or perhaps solely, from CNS projection neurons with tau pathology.

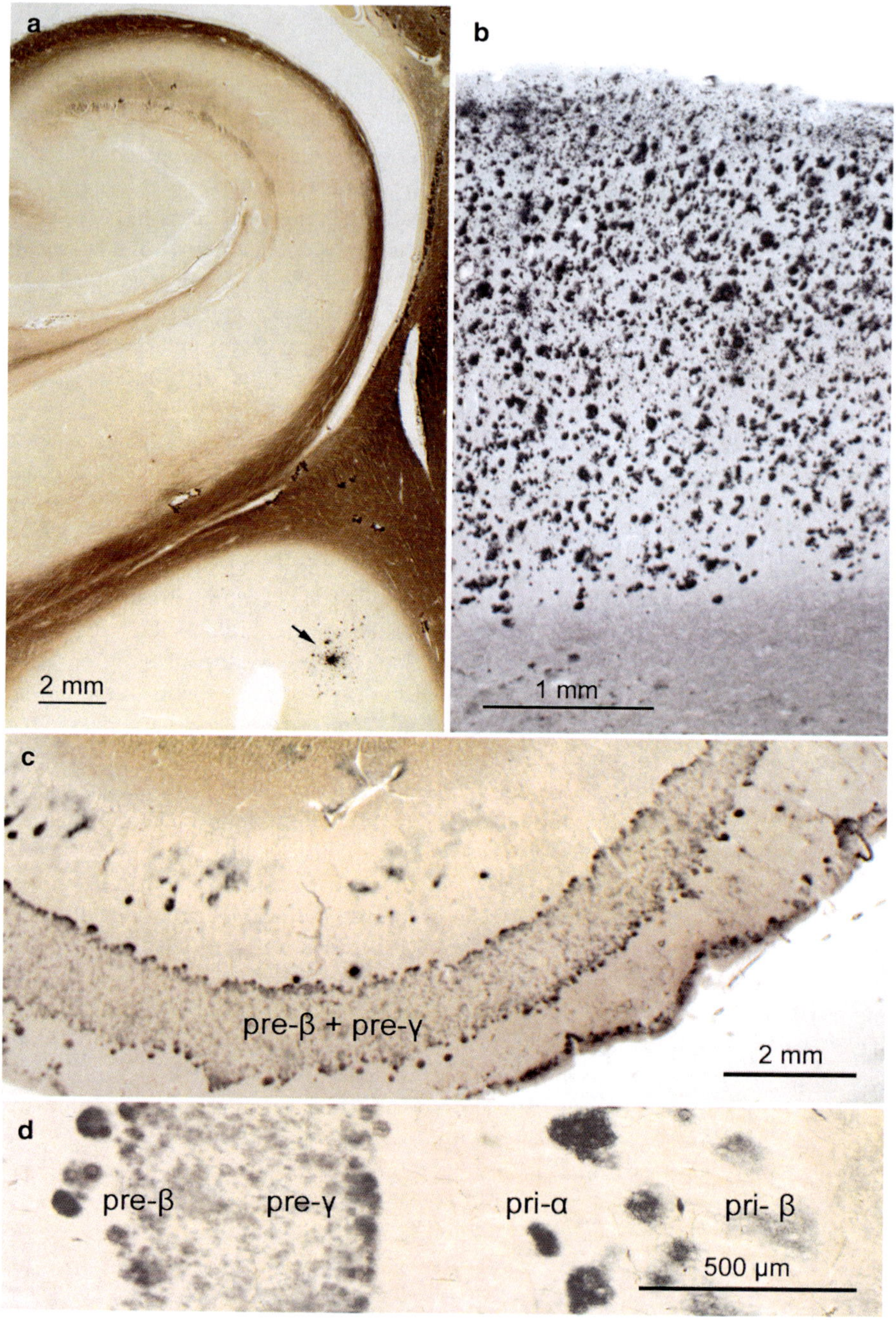

Fig. 8.2 Aβ plaques in 100 μm sections processed with the Campbell-Switzer silver-pyridine technique. (**a**) Phase 1: Initially, isolated plaques develop in the basal temporal neocortex (*arrow*) in the absence of plaques in the hippocampal formation (42-year-old male). (**b**) Phase 5: Maximal plaque density in the temporal neocortex of an 82-year-old demented male with AD (NFT stage V). (**c**) Band-like plaque formation in layers pre-β and pre-γ of the entorhinal region of in 87-year-

The fact that the somatodendritic domains of involved projection neurons are seldom surrounded by Aβ deposits raises the question at which cellular sites specifically (dendrites, soma, axon, synapses) Aβ is released into the ISF. Given what is already known about the typical plaque distribution pattern (Thal et al. 2002), it can be ruled out that Aβ is released via dendrites or cell bodies. In addition, it can be surmised that Aβ is not given off through most of the axonal membranes (for instance, at the nodes of Ranvier) because the white matter remains nearly devoid of Aβ deposition and only a few plaques are seen to develop near the cortical gray matter. Instead, Aβ deposits are more or less evenly distributed among the somatodendritic domains of nerve cells. Direct contacts with neurons occur only on a random basis and as a result of the high densities of both nerve cells and Aβ deposits (Fig. 8.2b). No direct evidence indicates a potential release of Aβ via the somatodendritic domain. Notably, some sites that harbor cell somata and dendritic processes with neurofibrillary changes, such as the locus coeruleus or layer pre-α of the entorhinal region, remain free of Aβ deposition (Fig. 8.2c). Involved coeruleus neurons have tau-immunoreactive inclusions in both dendrites and axons. However, whereas the axons extend into the cerebral cortex, which is richly supplied with Aβ plaques, the dendrites remain confined to the local neuropil of the brainstem, which contains very few plaques. Therefore, it is unlikely that Aβ is released from dendrites. Moreover, it has been shown that APP is transported along axons (Koo et al. 1990). This finding and the distribution pattern of Aβ plaques in general make it more likely that Aβ is released from presynapses of terminal axons, along which nerve cells normally release their neurotransmitter and/or neuromodulator substances (Stokin and Goldstein 2006; Muresan and Muresan 2008; Harris et al. 2010; Haass et al. 2012; Braak and Del Tredici 2013a).

In the course of the AD process, plaque-like Aβ deposits do not occur in the absence of intraneuronal tau pathology—they develop later than the tau lesions (Table 7.2; Fig. 9.16) (Silverman et al. 1997; Schönheit et al. 2004; Dong et al. 2012; Giacobini and Gold 2013; but see Hardy and Selkoe 2002; Price and Morris 2004; Hardy 2006; Golde et al. 2011; Karran et al. 2011; Mann and Hardy 2013). This means that Aβ deposition begins when specific types of nerve cells, e.g., nerve cells in the brainstem nuclei with diffuse cortical projections, already have undergone cytoskeletal tau changes. The assumption that Aβ is the initial causal event of the AD process is therefore erroneous (compare Fig. 9.16a and b) (Korczyn 2008; Pimplikar 2009; Duyckaerts 2011; Braak and Del Tredici 2013a, b; Jagust et al. 2012; Chételat 2013; Chételat and Fouquet 2013; Perani 2014).

Fig. 8.2 (continued) old male AD patient (NFT stage V), seen in greater detail in (**d**). Other amyloid precipitations, such as those occurring in prion diseases (spongiform encephalopathies), remain unstained. Fully developed silver-stained sections demonstrate a non-specific co-staining of axons. This readily and reliably applicable silver technique also distinctly demonstrates neuromelanin granules and Lewy bodies/neurites in Lewy body disease (PD) as well as argyrophilic oligodendrocytes associated with multisystem atrophy (MSA). See also the Technical addendum in Chap. 11.

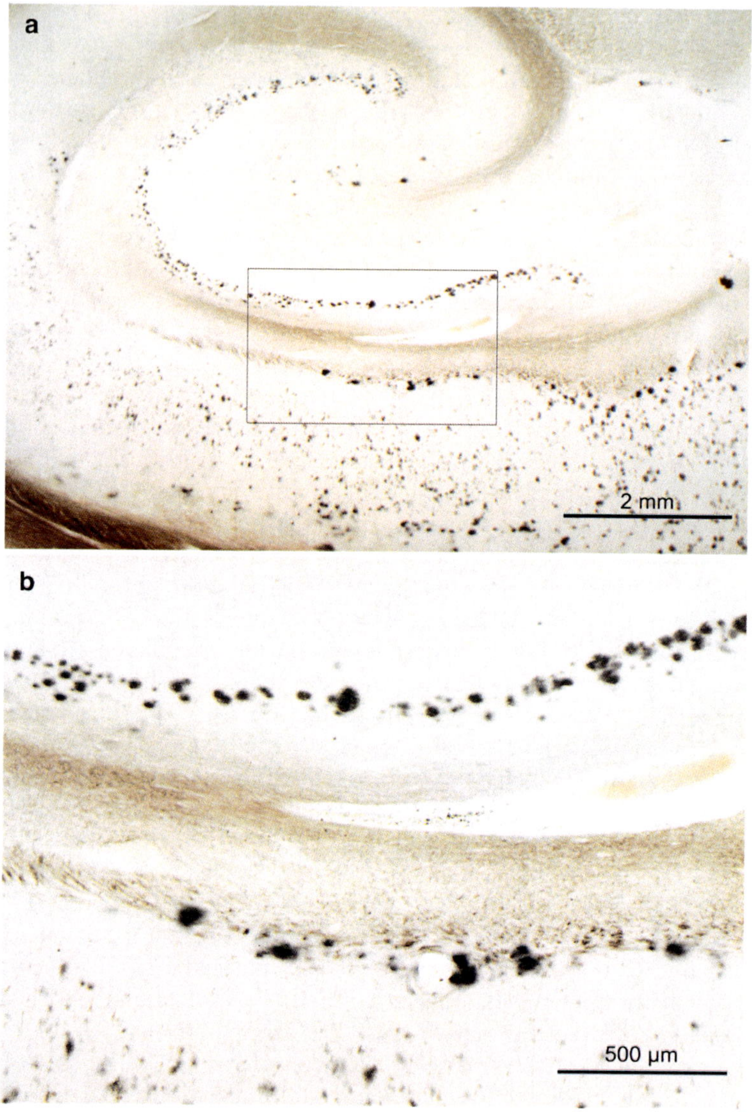

Fig. 8.3 Aβ plaques in 100 µm sections (Campbell-Switzer silver-pyridine method). (**a**) Phase 3: Aβ deposits develop in the hippocampal formation of a 67-year-old female. Note the densely packed row of plaques along the course of the perforant pathway not only in CA 1 but also in the molecular layer of the dentate fascia. (**b**) Higher magnification of the framed area in (**a**)

First, primitive (i.e., diffuse) Aβ plaques develop in the basal temporal neocortex (Braak and Braak 1991a) (Fig. 8.2a); in other words, at a time and in a region where pyramidal cells lack AD-associated tau aggregations. If our assumption is correct that Aβ only originates in nerve cells that are already involved in the AD process, then Aβ can only reach the basal temporal neocortex by way of long axons

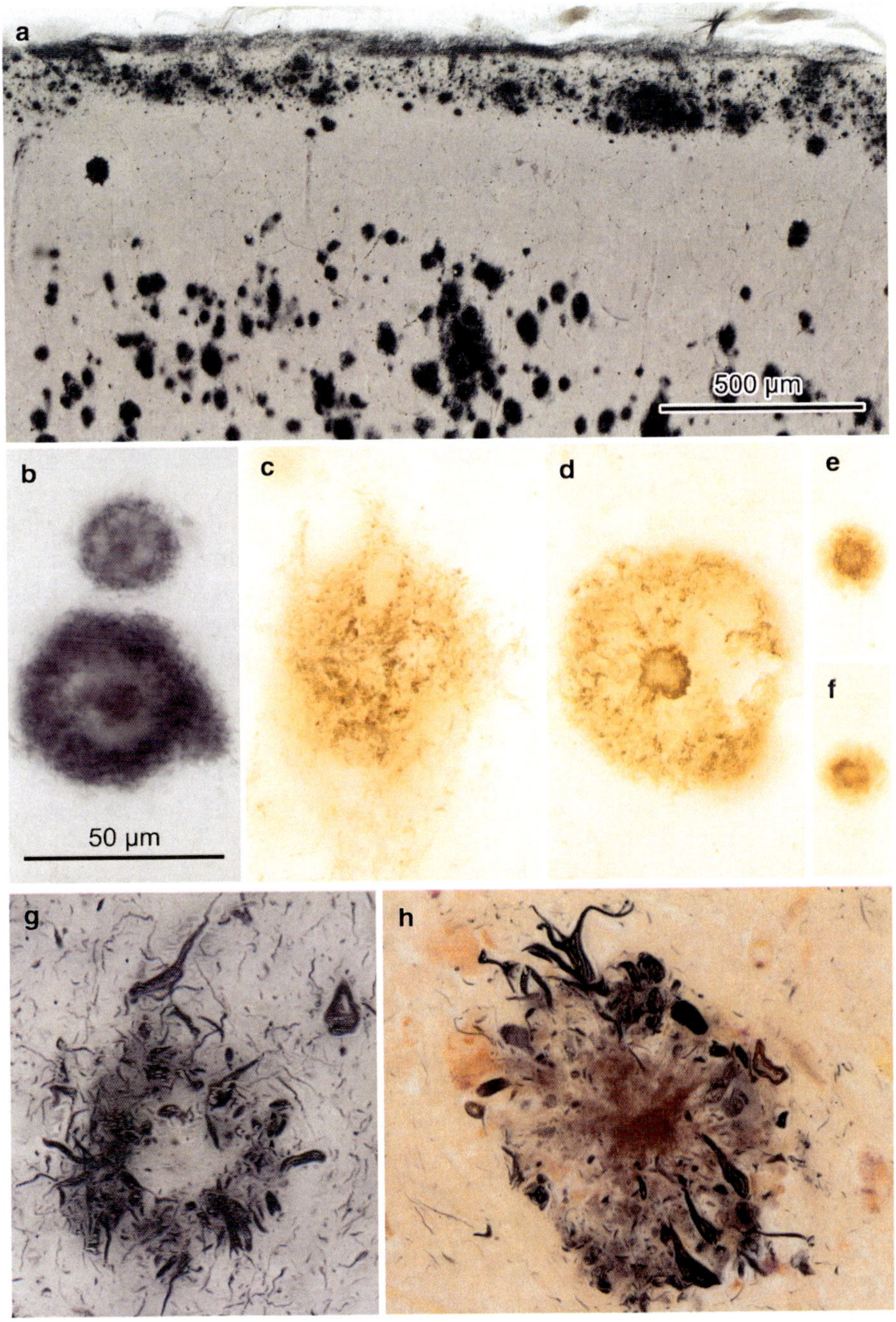

Fig. 8.4 Different forms of Aβ plaques in 100 μm sections. (**a**) Band-like deposits of Aβ directly subjacent the layer of surface astrocytic endfeet. Deeper portions of the molecular layer harbor densely packed globular plaques that frequently become confluent (female, 96 years of age, NFT stage VI). (**b**, **d**) Examples of cored plaques in an 84-year-old male (**b**, NFT stage V, Campbell-Switzer) and in a 60-year-old male (**d**, stage VI, 4G8 immunoreaction). (**c**) Diffuse plaques often

projecting to this part of the neocortex, a condition fulfilled by the axons of the diffusely projecting brainstem nuclei.

The existence of Aβ plaques in the cerebellum (Braak et al. 1989b; Thal et al. 2002) can best be explained by a similar phenomenon, i.e., the release of Aβ via terminal axons belonging to nerve cells with tau pathology, insofar as the various cerebellar neuronal types do not develop abnormal tau inclusions. They are, however, well supplied with a dense axonal network originating from brainstem nuclei, above all the locus coeruleus, where abnormal tau inclusions occur remarkably early.

In this context, it is necessary to reiterate that the terminal segment of the extensively branching axons of diffusely projecting brainstem nuclei develop large numbers of local thickenings with only presynaptic sites (so-called "non-junctional varicosities") in the absence of postsynaptic counterparts. By means of these varicosities, they release their neurotransmitter and neuromodulator substances (volume transmission) diffusely into the ISF (Agnati et al. 1995; Nieuwenhuys 1999; O'Donnell et al. 2012). It is conceivable that soluble forms of Aβ may likewise be released at non-junctional varicosities directly into the ISF (Braak and Del Tredici 2013a). This interpretation is supported by the existence of Aβ deposits that are found around the smooth muscle layer of vessel walls in the CNS in the form of cerebral amyloid angiopathy (CAA) (Yamada and Naiki 2012) (see Sect. 8.7). Moreover, since axons of the diffusely projecting brainstem nuclei only spread throughout the CNS—a volume transmission mechanism would also account for why Aβ plaque formation remains confined to the CNS and does not develop in the PNS and ENS (for the olfactory mucosa, however, see Arnold et al. 2010).

With the notable exception of Aβ plaques in the striatum the dense network of coeruleus noradrenergic terminals corresponds remarkably well to the topographic distribution pattern of both Aβ plaques and CAA in sporadic AD (Counts and Mufson 2012). It still must be shown whether Aβ is preferentially given off from terminals of coeruleus neurons and whether additional nuclei with diffuse projections also contribute to the production of Aβ plaques, such as the terminals of the upper raphe nuclei, which, in turn, could explain the development of Aβ plaques in the striatum (Braak and Del Tredici 2013a).

The pallidum is an expansive forebrain region that is not reached by ascending projections originating from noradrenergic, serotonergic, or cholinergic non-thalamic nuclei. This fact accounts for the previously mentioned and puzzling

---

Fig. 8.4 (continued) show ill-defined surfaces (same stage VI case as in **d**, 4G8 immunoreaction), whereas cored plaques (**d**) mostly have clear-cut outlines. (**e**, **f**) Burned out plaques are much smaller and generally have a core (same individual as in **d**, 4G8 immunoreaction). (**g**, **h**) Examples of NPs in a 74-year-old male (**g**) and in a 60-year-old male (**h**). Gallyas silver-iodide impregnations stain a network of argyrophilic neuronal processes in peripheral portions of amyloid deposits. The amyloid core is unstained in (**g**) and differently stained in a violet shade in (**h**). Scale bar in (**b**) applies also to (**c**–**h**)

finding that both segments of the pallidum belong to the very few regions of the forebrain that do *not* develop Aβ plaques. Unclear is whether a similar relationship can also be found for the absence of Aβ deposits in selected regions of the hypothalamus (i.e., the lateral tuberal nucleus and lateral mamillary nucleus).

The perforant pathway also deserves mention because it is frequently decorated with Aβ deposits (Fig. 8.3). Projection cells in the external entorhinal cellular layers give rise to this glutamatergic path that terminates in the hippocampal formation (CA 1 and dentate fascia) (Hyman et al. 1988; Braak et al. 1996). The host entorhinal neurons tend to develop intraneuronal tau inclusions early, and Aβ deposits are often found later close to the terminal ramifications of their axons. Perforant pathway fibers contact only a portion of the dendritic tree of CA 1 projection neurons, whereas dendritic segments outside of the pathway are initially free of Aβ deposits. For these reasons, axon terminals of the perforant path may also be capable of releasing Aβ (Fig. 8.3) (Buxbaum et al. 1998; Harris et al. 2010). It is still not known whether axons of the perforant path are endowed with non-junctional varicosities.

In the ISF, the hydrophobic but still soluble Aβ molecules are prone to further aggregation that may be induced by seeding sites. The pathological material ultimately converts into insoluble plaque-like deposits of variable sizes and shapes (Figs. 8.2–8.5) (Thal et al. 2002). Insoluble Aβ precipitations can be visualized using Campbell-Switzer silver-pyridine staining or immunoreactions (Campbell et al. 1987; Braak and Braak 1991b; Montine et al. 2012). The aggregated amyloid fibrils of primitive or cored plaques in the cortex are rich in cross-β sheet structures (Haass et al. 2012; Masters and Selkoe 2012), whereas these components in the non-amyloid Aβ plaques of the striatum and cerebellum are sparse. Because aggregated fibrils possess low bioactivity, we are inclined to see them as posing no immediate danger to adjoining components of the neuropil. From then on, the potentially undesirable side-effects of the Aβ plaques essentially would arise from their capacity to displace other structures. In fact, given the limited dimensions of the extracellular space in the CNS and its importance for the functionality of nerve cells, it is certainly conceivable that such side-effects could occur.

Once Aβ production begins, the total volume of insoluble Aβ plaques increases noticeably, and their number steadily increases until, apparently, a certain level is reached. Inasmuch as Aβ production continues for decades and (if at all) degradation of plaques only occurs slowly, one would expect plaques to eventually fill the entire cortical gray matter. However, in the end phase of AD, notable portions of the gray matter still are devoid of Aβ deposits even in cortical regions that are heavily laden with plaques (Fig. 8.2b). In other words, it looks as if, once a maximal plaque density has been attained, this status remains unchanged for a protracted period of time. Factors mediating the gradual reduction and final cessation of Aβ production (Hyman et al. 1993) may include the impairment and failure, over decades, of projection neurons in the non-thalamic nuclei with diffuse cortical projections. The growing presence of tombstone tangles in these nuclei would be a sign of the lost numbers of axons capable of generating Aβ.

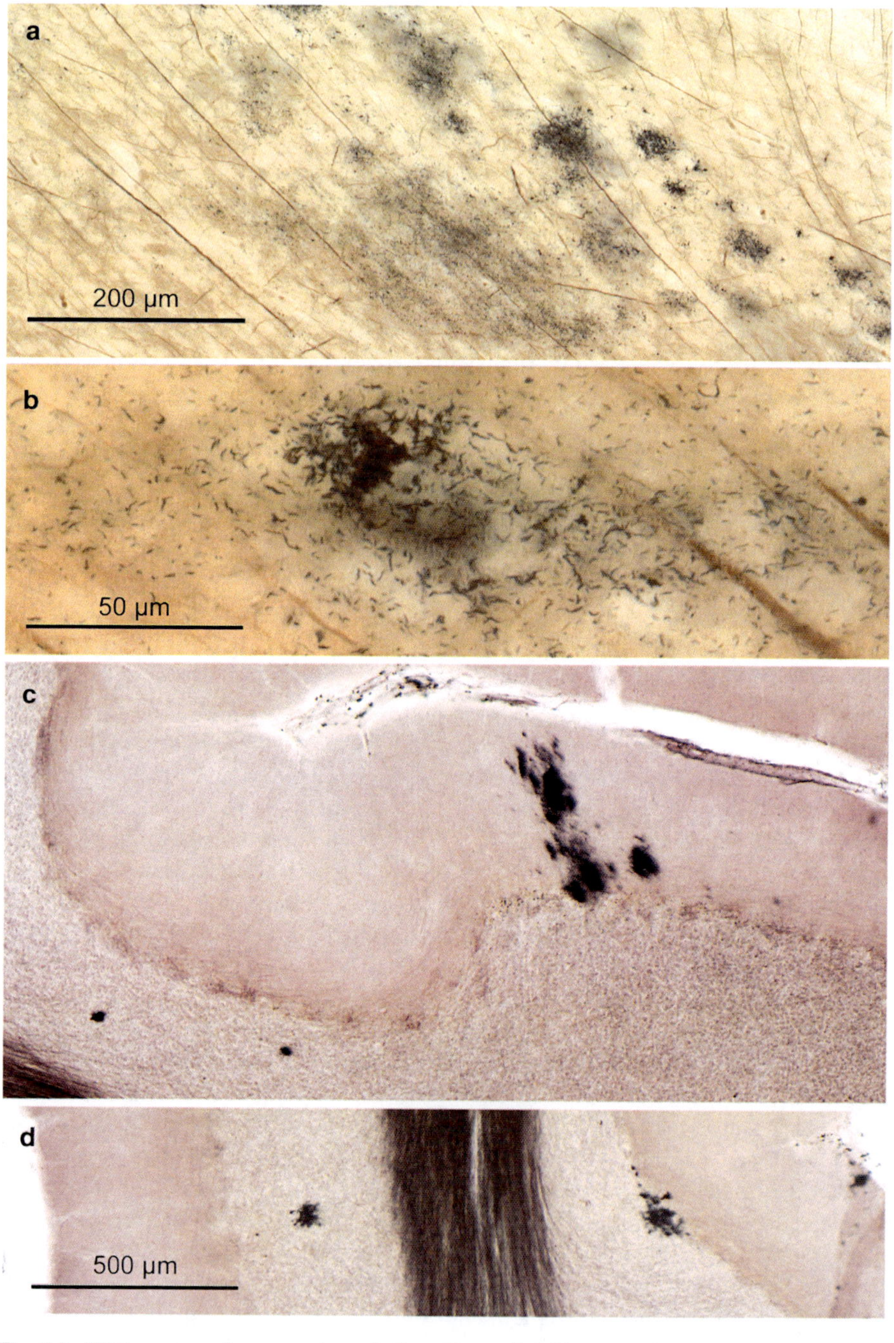

Fig. 8.5 White matter plaques and cerebellar plaques in 100 μm sections. (**a**, **b**) White matter plaques usually are located close to the cortical gray matter and consist of irregularly shaped and only weakly stained flake-like deposits (**a**), which gradually condense into more compact forms, as seen in greater detail in (**b**) (68-year-old female, NFT stage III). (**c**, **d**) In phase 5, the cerebellum develops non-amyloid Aβ in the form of globules of various sizes in the granular layer (**c** *left side*)

Toxic effects on the surrounding neuropil and vessel walls are attributable primarily to Aβ when the peptide is still in a soluble and diffusible (oligomeric) form (Mucke and Selkoe 2012). However, such forms frequently evade detection in conventionally fixed tissue. Soluble forms possibly can bind to membranes not only of nerve cells but also of non-neuronal cells (Mucke and Selkoe 2012). Cellular processes from neurons and glia in the immediate vicinity of plaques can develop dystrophic processes, including accumulations of dense bodies and abnormal mitochondria. During the transition from primitive to neuritic plaques, cellular processes may even develop abnormal tau inclusions (i.e., argyrophilic dystrophic neurites). In addition, the synapses located near Aβ plaques may display signs of deterioration (Fiala 2007).

Some of the soluble and diffusible Aβ released into the ISF reaches the narrow space between the capillary wall and the end-feet of astrocytes, and it passes from there through gaps between smooth muscle cells of the vessel walls and the enveloping glia sheath. In this manner, the pathological protein drains into the regional lymph nodes of the neck similar to drainage of lymphatic fluid (Weller et al. 1998, 2009).

8.3 Transient Extracellular Aβ Deposits

At the outset of the phase when accumulations of Aβ protein that have precipitated out of the formalin fixative gradually develop out of soluble and diffusible Aβ monomers and oligomers, inconspicuous and, in part, very extensive cloud-like Aβ deposits with blurred boundaries emerge temporarily. Such deposits usually are observed in deep layers of the cortex, for instance in layer VIb of the temporal neocortex and in the deep entorhinal layer pri-γ (Fig. 8.2c, d). Faintly tinged Aβ strands develop in these layers and widely infiltrate the tissue, tending to merge into each other, and often extend into the white substance (Thal et al. 1999). There, the material gradually becomes compressed into tightly packed granules, i.e., white matter plaques (Fig. 8.5a, b) (Braak et al. 1989b).

Because of the transient nature of the cloud-like Aβ deposits, there is no readily quantifiable relationship between the depth of Aβ infiltration into the white substance and disease duration. Transient forms of Aβ deposition are congo red-negative, a fact indicating that the fibrils in the tissue are not yet cross-β sheet-rich assemblages. These transient morphological manifestations indicate that Aβ, immediately after its release into the ISF, aggregates with other Aβ molecules. The low viscosity of these formations prevents their equilibration and causes localized

Fig. 8.5 (continued) and Purkinje cell layer (**d** *right side*) or as rectangular slices (**c**) along dendritic trees of Purkinje cells in the molecular layer (77-year-old male, stage III). Scale bar in (**d**) applies to (**c**)

differences in Aβ concentrations, especially between the ISF and CSF (Englund et al. 2009).

Similarly uniform band-like Aβ formations often develop, although not inevitably, in pyramidal cell layers of CA 1, and in entorhinal layers pre-β and pre-γ (Fig. 8.2c, d). By contrast, layer pre-α generally remains free of deposits (Mufson et al. 1999). Such band-like deposits have not been reported in cases of fully developed AD and thus likely represent an intermediate type between the inconspicuous extensive cloud-like formations and the permanent spherical Aβ deposits. Only the uniform lake-like depositions of Aβ in the parvocellular layer of the presubiculum remain at this location up to the end-phase of AD (Kalus et al. 1989). The mechanisms by which the Aβ deposits develop from expansive and unsharply defined infiltrations to compact globular plaques are not known. It is possible that the migration of glial cells contribute somehow to the morphological changes undergone by Aβ deposits.

8.4 Mature Forms of Aβ Deposits and Plaque Degradation

Sharply outlined globular amyloid deposits of varying diameters represent mature forms of either primitive or cored Aβ plaques (Figs. 8.2b and 8.4b–d). They are composed of extracellular wisps of amyloid braided with assemblies of swollen dystrophic nerve cell processes (Dickson 1997b; Tolnay and Probst 1999). In contrast to primitive (diffuse) plaques (Fig. 8.4b), cored plaques possess a central amyloid mass, which is often enclosed by microglial cells and astrocytes (Fig. 8.4c). Such cores can also occur as isolated structures and in many cases are referred to as "burnt out" plaques (Fig. 8.4e, f) (Dickson 1997b; Tolnay and Probst 1999; Dickson and Vickers 2001; Fiala 2007). In contrast to transient Aβ deposits, globular plaques rarely fuse. The aggregation-prone Aβ42 is found predominantly in the core of such plaques. Dot-like initiation sites (seeds) probably lead to self-aggregation of the material and to the radial orientation of the filaments. Cored amyloid deposits can easily be mistaken for neuritic plaques (NPs) but differ from NPs by the absence of argyrophilic neuronal processes (Fig. 8.4b, d).

The various layers of the cerebral cortex display idiosyncracies locally. For example, subpial portions of layer I frequently contain confluent plaques (Fig. 8.4a), whereas layers II, IV, and Vb/VI are spared or contain only a few deposits. Spherical deposits predominate in neocortical layers III and Va. Dots are a specific feature of layer IVcα in the primary visual field (Fig. 9.8c) (Braak et al. 1989b). A few amyloid deposits also routinely occur in the white matter close to the transition to the cortex (i.e., white matter plaques) (Fig. 8.5a, b). Usually, no amyloid deposition is found in deeper portions of the white matter. Even the fiber tracts running through Aβ-laden portions of the gray matter (e.g., fornix, mamillothalamic tract, anterior commissure) only display a few precipitations. The lone exception is the perforant path that is densely decorated with Aβ plaques (Fig. 8.3) (Fiala 2007). The material in primitive plaques originally is a malleable mass that can adapt its shape to fit any structure of a given neuropil. Thus, globular Aβ deposits arise in the granular layer of the cerebellar cortex (Fig. 8.5d), while plaques in the molecular layer tend to extend vertically so as to follow the dendritic trees of Purkinje cells (Fig. 8.5c) (Braak et al. 1989b).

In the human brain, an inverse relationship exists between the degree of cortical myelination and the density of Aβ deposition. Sparsely myelinated cortical areas and layers display denser deposits than those that are rich in myelin. Densely myelinated layers, such as neocortical layers IV and Vb, which harbor the outer and inner lines of Baillarger, as well as the myelin-rich molecular layer of the allocortex remain free of Aβ deposits or show only few plaques. Areas and layers containing pyramidal cells that are rich in lipofuscin deposits also tend to show denser accumulations of globular Aβ deposits than those with sparsely pigmented neurons.

8.5 Phases in the Development of Aβ Deposits

Changes in the regional distribution pattern of plaques are less predictable than those of the intraneuronal tau inclusions, however, the gradual deposition of Aβ plaques also follows a stereotypic five-phase pattern as shown in cross-sectionally studied cases (Fig. 8.6) (Thal et al. 2002).

Initial Aβ deposits (phase 1) typically develop in the poorly myelinated basal portions of the temporal and frontal neocortex, mostly in the form of diffuse plaques dispersed throughout the richly pigmented layers III and Va and in portions of the cortex located in the depth of sulci (Fig. 8.2a) (Braak and Braak 1991a; Thal et al. 2002). During phase 2, band-like deposits of Aβ begin to develop in the entorhinal layers pre-β, pre-γ (Fig. 8.2c), and in the first sector of the Ammon's horn. Sometimes, additional plaques also develop in the amygdala, as well as in the insular and cingulate cortex. Thereafter, plaque formation takes place in the molecular layer of the dentate gyrus together (Fig. 8.3) with the deposition of lake-like Aβ in the presubicular parvocellular layer (phase 3). Diffuse plaques in the basal temporal neocortex rapidly increase in number and then extend into adjoining neocortical association areas, temporarily sparing only the belt regions and the primary fields. Plaques also develop in subcortical sites in phase 3, such as the striatum, magnocellular nuclei of the basal forebrain, and the gray matter of both the thalamus and hypothalamus. Phase 4 is marked by the expansion of Aβ deposition into the fourth sector of the Ammon's horn as well as into virtually all areas of the neocortex. Involved regions in the lower brainstem during this phase include the red nucleus, as well as the superior and, in particular, the inferior colliculi of the mesencephalon. During the final phase 5, Aβ deposition reaches the reticular formation of the lower brainstem and, notably, the cerebellar cortex (Figs. 8.5c, d and 8.6) (Thal et al. 2002).

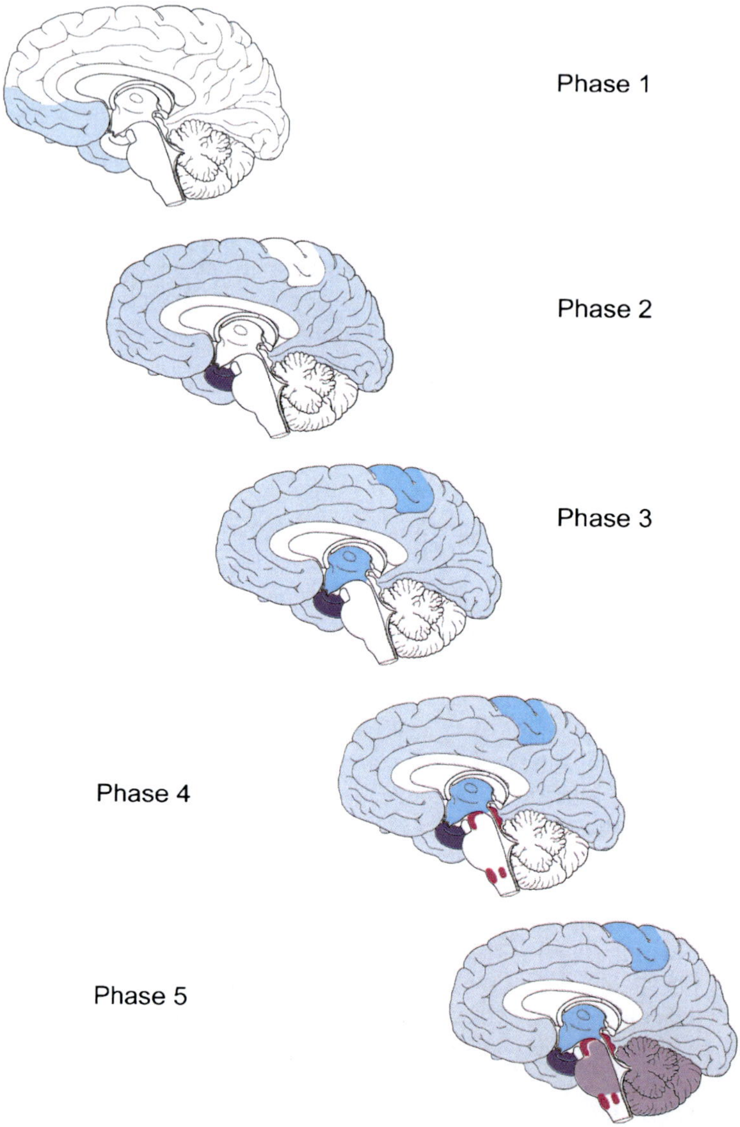

Fig. 8.6 Phases 1–5 in the development and progression of Aβ deposits. The regional distribution pattern is shown by different degrees of shading for each stage (*light blue*, *dark blue*, *turquoise*, *magenta*, *purple*). In phase 1, isolated plaques develop at one or more sites within the basal temporal and the orbitofrontal neocortex. Additional plaques are found in phase 2 in the allocortex and amygdala. Plaques develop in virtually all high order association areas of the neocortex. Phase 3 is marked by further expansion of Aβ plaques into secondary neocortical fields and into the striatum. They also appear in the perforant pathway and presubiculum. In phase 4, plaque formation is seen in virtually all areas of the neocortex and reaches the mesencephalon, particularly the inferior colliculi. During the final phase 5, Aβ deposition reaches the lower brainstem and cerebellar cortex

8.6 Formation of Neuritic Plaques (NPs)

NPs consist of abnormal astrocytes, microglial cells, dystrophic neuronal processes, neuronal processes filled with AT8-ir non-argyrophilic material, and cellular processes filled with argyrophilic (Gallyas-positive) tau aggregates (Fig. 8.4g, h). Aβ deposits accompany NPs in the form of peripheral infiltrations and, frequently, compact cores (Fig. 8.4h). NPs are fewer in number and less widely distributed than primitive or cored Aβ plaques. Accurate assessment and the distinction of NPs *versus* diffuse Aβ deposits is essential for the CERAD-based diagnosis of the pathological process underlying AD (Mirra et al. 1991; Hyman and Trojanowski 1997; Montine et al. 2012).

Some regions (striatum, cerebellum) of the CNS remain devoid of NPs. The non-amyloid Aβ plaques that occur there do not convert into NPs. Thus, it can be conjectured that a sufficient density of cross-β sheet structures is needed for the conversion of primitive plaques into NPs. Other regions, by contrast, such as the transentorhinal region and the primary visual cortex, display high densities of NPs quite early. The cortex covering the depths of the sulci generally exhibit a higher density of NPs that that covering the crests of the gyri.

NP production begins only when large expanses of the neocortical neuropil have already been permeated with primitive Aβ plaques, i.e., NPs are absent in early disease stages (Braak and Braak 1991a). Under such conditions, it is conceivable that terminal ramifications of candidate axons, e.g., axons of non-thalamic nuclei projecting to the cerebral cortex that already contain non-fibrillar abnormal tau, may do something they would not do otherwise, namely, convert (possibly influenced by Aβ?) their non-argyrophilic material into Gallyas-positive fibrils, i.e., conversion into the argyrophilic material typically found in dystrophic neurites of NPs. Some of these dystrophic axons also contain markers for cholinergic, gabaergic, and glutamatergic transmission (Kitt et al. 1985a, b; Struble et al. 1985; Whitehouse et al. 1985), and it is possible that such dystrophic axons are formed through a plaque-induced axonal sprouting process (Masliah et al. 2003). Tombstone tangles occasionally can serve as additional nucleation sites of Aβ material and may similarly attract axons of diffusely projecting brainstem nuclei and eventually lead to the formation of tangle-associated neuritic clusters (TANCS) (Munoz and Wang 1992).

8.7 Cerebral Amyloid Angiopathy

Precipitations of Aβ mainly consisting of Aβ40, which is less prone to aggregation than Aβ42, frequently are seen in close association with the abluminal basement membrane and between muscle cells of the *tunica media* in leptomeningeal and cortical vessels (Elfenbein et al. 2007; Revesz et al. 2009; Viswanathan and Greenberg 2011; Grinberg et al. 2012; Yamada and Naiki 2012; Love et al. 2014). The usual designation for this pathological material is cerebral amyloid angiopathy (CAA), which involves arteries, arterioles, and, less frequently, capillaries (Attems et al. 2011; Love et al. 2014). In the cerebral cortex, CAA generally and inconspicuously begins in middle portions of the molecular layer (Fig. 8.7a, arrows). From there, it extends into both the cortical gray and the leptomeninges (Fig. 8.7a, b). CAA predilection sites do not diverge substantially from those of cortical Aβ plaques. The vasculature of the occipital and parietal lobes often is more severely involved than that of the temporal lobe and other cortical regions. CAA accompanies the AD-associated process: About 80 % of cases with clinical AD have concurrent CAA, mostly in a mild form.

The origin of CAA is controversially discussed. Soluble forms of Aβ in blood plasma and in smooth muscle cells of vessel walls have both been held responsible for CAA production (Frackowiac et al. 1994; Mackic et al. 2002). Nonetheless, the current consensus is that the Aβ precipitation in the vasculature originates chiefly in neurons. We have suggested (see Sect. 8.2) that soluble and diffusible forms of Aβ enters the ISF by means of volume transmission from abnormal tau-containing neurons of the locus coeruleus and other non-thalamic nuclei with diffuse ascending projections (Braak and Del Tredici 2013a, b). The smooth muscle cells of the vasculature may be endowed with seeds that can prompt the material released into the ISF to form aggregates. CAA partially marks a drainage pathway for the ISF that, beginning at the glial top layer of the Virchow Robin space, arrives at the regional lymph nodes of the neck via the vascular sheaths of the leptomingeal and extracerebral arteries (Fig. 8.7c, d) (Weller et al. 1998, 2009). The question why some portions of the cerebrovascular system display CAA especially early on in the AD process, whereas others show very severe forms, remains unanswered.

Three stages in the evolution of CAA have been proposed (Thal et al. 2003). CAA initially involves vessels in the leptomeninges and neocortex (stage 1). In stage 2, these lesions extend into the allocortex (hippocampus, entorhinal region) and are seen in cortical transition areas (transentorhinal region, insular, cingulate proneocortex), subcortical components of the limbic system (i.e., amygdala), and occasionally in the hypothalamus and cerebellum. The third and final stage includes CAA in vessels of the basal ganglia, thalamus, lower brain stem, and cortical white matter (Thal et al. 2003).

CAA often fails to result in clinically recognizable symptoms and usually is diagnosed only at autopsy; however, severe CAA can cause major intracranial hemorrhages. Moreover, the presence of microbleeds or lobar hemorrhages (predominantly in the occipital lobe) in demented individuals with AD frequently is attributable to the presence of concomittant severe CAA (Vonsattel et al. 1991).

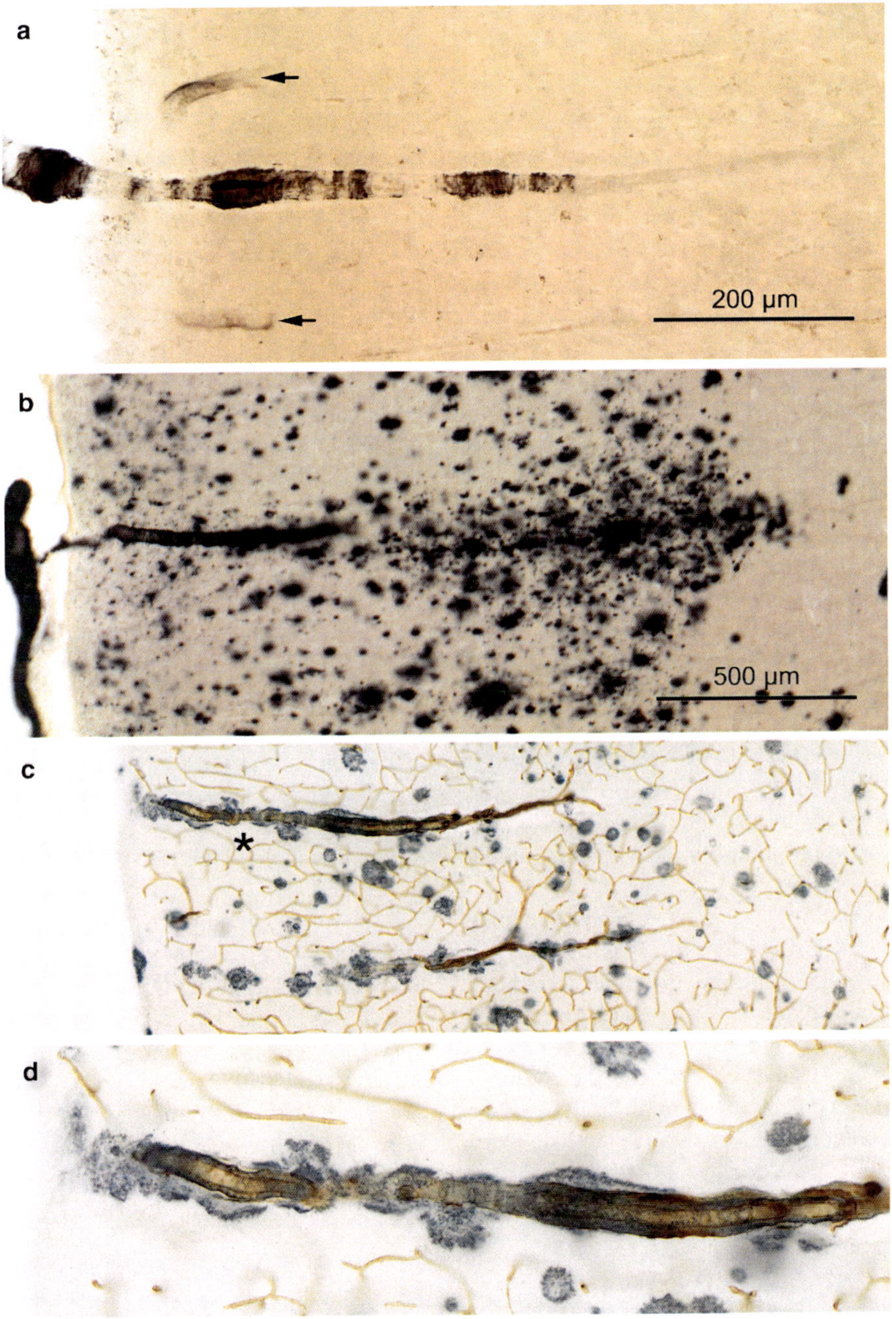

Fig. 8.7 Cerebral amyloid angiopathy (CAA) in 100 µm sections. (**a**) Depositions of Aβ occur close to the abluminal basement membrane and to muscle cells of leptomeningeal (*left*) and cortical arteries. This development notably begins in middle and lower portions of the molecular layer. *Arrows* point to initial changes. The lesions can develop in the absence of free cortical Aβ plaques (85-year-old male, NFT stage II, Campbell-Switzer silver-pyridine method). (**b**) Involved

8.8 Soluble Aβ as a Biomarker in the CSF

Neuronally generated soluble and diffusible Aβ may be given off into the ISF (Sect. 8.1). From there, equilibrated Aβ may be partially drained off via perivascular spaces into the extracranial lymph nodes or enter (albeit in considerably lowered concentrations) the CSF. Changes within the CSF may be able to provide information about pathological processes in the CNS but only when an intact and free exchange between the ISF and CSF is possible (Blennow et al. 2010). Soluble Aβ forms are consistently found in the CSF, and Aβ levels are thought to reflect a steady state equilibrium between the processes of Aβ production and clearance (Karran et al. 2011). Nevertheless, the inter-individually comparable and long-term more or less stable Aβ level in CSF is difficult to explain because not only is the Aβ level in a state of dynamic equilibrium but the CSF is also constantly produced and re-absorbed. Furthermore, it is unknown whether the concentration gradient between the ISF and CSF remains constant throughout the course of the AD-process, or whether it is subject to fluctuations.

As pointed out above in Sect. 8.2, it can be assumed that Aβ is not produced by healthy nerve cells but, rather, is released chiefly into the ISF via volume transmission by axons belonging to previously involved neurons. The resulting amount of Aβ is quantitatively marginal and distributed widely throughout the CNS. Given its low concentration, it is unlikely that Aβ molecules interreact with each other, and the result is that soluble Aβ reaches a plateau in the CSF in initial phases of the disease process (stages a-II). Subsequently, when threshold values are exceeded, localized Aβ-plaque formation begins (sometimes in stages 1a,1b but more frequently during the transition from stage I to stage III; see Table 7.2). Among the consequences is an approximately 50 % reduction of the CSF Aβ level in comparison to that found in non-demented controls (Blennow and Hampel 2003; Englund et al. 2009).

Additional data have become available from the development of Aβ ligands for use as positron emission tomography (PET) tracers to visualize fibril Aβ deposits in the brain of living individuals (Klunk et al. 2004; Jack 2012). It has been noted that striatal plaques correspond to higher NFT stages and, thus, that striatal amyloid imaging could be a predictor for symptomatic AD (Beach et al. 2012b). An inverse relationship exists between the Aβ plaque load assessed by PET tracers and that of CSF Aβ levels (Fagan et al. 2009; Ikonomovic et al. 2008; Grimmer et al. 2009; however, see also Kepe et al. 2013), and CSF levels of Aβ inversely correlate with

---

Fig. 8.7 (continued) arteries are frequently surrounded by particularly densely packed and small cortical Aβ deposits (76-year-old female, NFT stage III, Campbell Switzer silver-pyridine method). (**c**, **d**) More or less equally distributed Aβ deposits along astrocytic endfeet of the superficial glial layer bordering the space of Virchow Robin (71-year-old male, stage VI, Campbell Switzer silver-pyridine method combined with collagen IV immunoreaction). *Asterisk* (*) in (**c**) indicates a vessel shown in greater detail in (**d**). Scale bar in (**a**) is valid for (**d**), scale bar in (**b**) applies also to (**c**)

the neuropathologically assessed plaque load seen at autopsy (Tapiola et al. 2009). Once Aβ plaques develop, however, they rapidly increase in number and, when a peak density is reached, additional plaque generation is either reduced or even ceases during the final disease stages. Current studies of Aβ-CSF levels do not reflect this gradual diminution and cessation of Aβ production.

Decreased Aβ concentrations combined with increased CSF levels of tau are referred to as the 'AD CSF profile' or 'AD CSF pattern' (Frankfort et al. 2008; Jack 2012; Jack et al. 2013; Rosén and Zetterberg 2013), and such a profile usually evolves before the clinical symptoms of AD can be diagnosed. Reduced CSF Aβ levels owing to plaque formation generally precedes increases in CSF T-tau (Li et al. 2007; Fagan et al. 2009). However, this sequence of CSF biomarker events does not correspond to that in which abnormal tau and Aβ appear in the brain (see Sect. 8.2), where the formation of intraneuronal tau aggregates precedes the production of extracellular Aβ-plaques (Braak et al. 2013). Moreover, longitudinal studies reveal that the AD CSF profile remains stable between phases of mild cognitive impairment and those of overt dementia (Mattsson et al. 2012; Rosa et al. 2014).

Chapter 9
The Pattern of Lesions During the Transition to the Symptomatic Phase and in Fully Developed Alzheimer's Disease

9.1 NFT Stage III: Progression into the Basal Temporal Neocortex, Including Portions of the Fusiform and Lingual Gyri, Involvement of Superordinate Olfactory Centers and the Limbic Thalamus

The pathology seen at stage III still is concentrated chiefly in select allocortical regions and related transition areas. From there, it encroaches only to a limited extent upon the adjoining mature temporal neocortex (Figs. 9.2a and 9.13). The tau lesions that were present during NFT stage II worsen (Fig. 9.1g–i), and subcortical lesions occur in all brainstem nuclei with diffuse cortical projections. This tau pathology slowly increases and reaches its greatest extent in stage VI. In stage III, AD-associated tau lesions occur in small numbers of α-motoneurons of the spinal cord ventral horn (Fig. 9.6b) (Dugger et al. 2013). In addition, the lesions appear in the brainstem nuclei that regulate the extrinsic muscles of the eyes, above all the rostral interstitial nucleus of the medial longitudinal fascicle, and, to a somewhat lesser extent, the Edinger Westphal nucleus, nucleus of Darkschewitsch, and interstitial nucleus of Cajal. As a result, it would come as no surprise if the involvement of these regions were to be accompanied clinically by a slowing of vertical saccades (Rüb et al. 2001a).

These developments are accompanied by the appearance of the first abnormal tau aggregates in anterosuperior portions of the reticulate nucleus of the thalamus as well as in limbic subnuclei of the thalamus (reuniens nucleus and anterodorsal nucleus) (van der Werf and Groenewegen 2002). In addition, with the beginning involvement of the central subnucleus of the amygdala, the high order processing nuclei of the autonomic system are drawn into the pathological process. Further brainstem nuclei that display prominent changes already during stage III include the mesencephalic medial and lateral parabrachial nuclei, the subpeduncular nucleus,

H. Braak, K. Del Tredici, *Neuroanatomy and Pathology of Sporadic Alzheimer's Disease*, Advances in Anatomy, Embryology and Cell Biology 215,
DOI 10.1007/978-3-319-12679-1_9

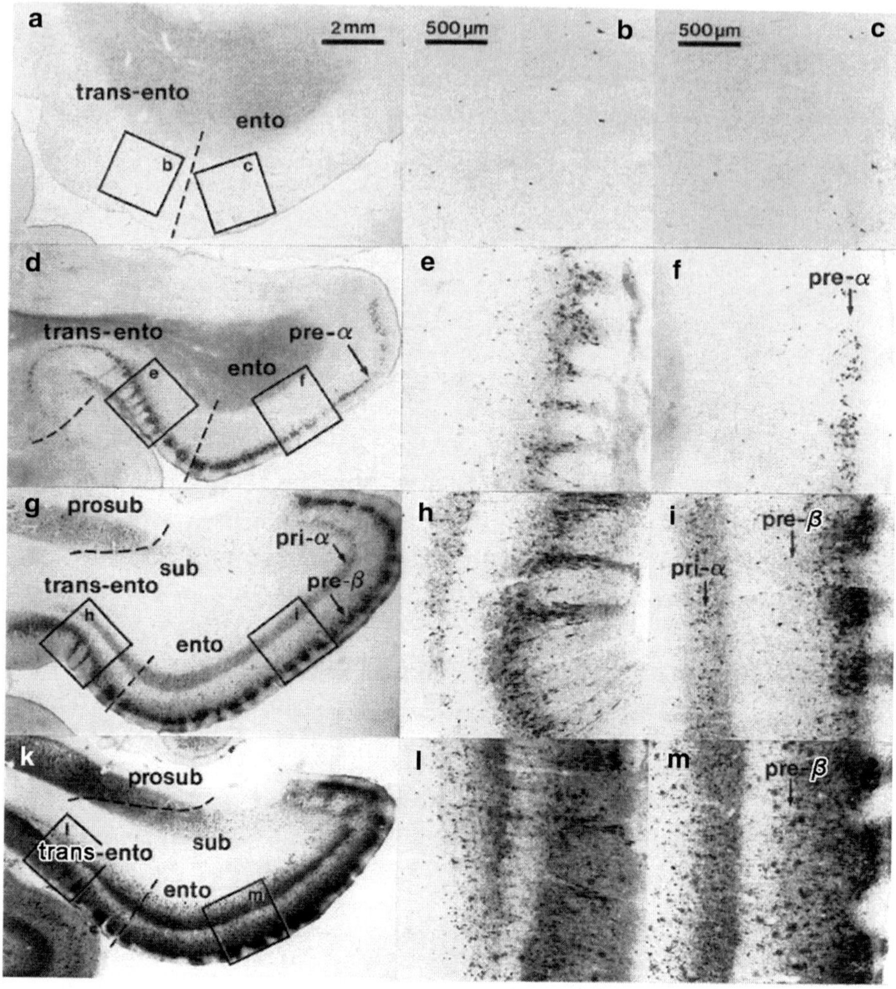

Fig. 9.1 Gradual progress of destruction of both transentorhinal and entorhinal regions during NFT stages I–IV in 100 µm sections (Gallyas silver-iodide technique). The insets provide overviews, with framed areas indicating the locations of the micrographs shown at the *right* at higher magnification. At the *left*, portions of the transentorhinal regions (trans-ento) and entorhinal regions (ento) appear with the pial surface oriented downward. These are supplemented by portions of the transentorhinal (**b**, **e**, **h**, **l**) and entorhinal (**c**, **f**, **i**, **m**) regions with the pial facing toward the right at higher magnification. Reproduced with permission from H Braak and E Braak, Temporal sequence of Alzheimer's disease-related pathology. Cerebral Cortex 1999;14:475–512

and the intermediate zone of the medullary reticular formation (Rüb et al. 2001b). At each of these sites, the sequence is the same: First, AT8-ir lesions develop, and these subsequently convert into argyrophilic NT/NFTs.

In stage III, the lesions expand into superordinate components of the olfactory system, including the olfactory tract, piriform and periamygdalear areas, and olfactory portions of the amygdala and entorhinal region (Christen-Zaech et al. 2003;

Kovács 2013). The outer cellular layers of the entorhinal region (pre-α, pre-β, pre-γ) become filled with a mesh of AT8-ir neurites (Fig. 9.4a, b), whereas the pale *lamina dissecans* exhibits only a few radially oriented neuronal processes. The striking devastation of pre-α projection cells is a key feature of stage III (Fig. 9.2a, b). Isolated neurons in these layers die at this early stage and leave behind tombstone tangles (Fig. 9.4g, h). The deep layer pri-α also is heavily involved and gradually thins within the transentorhinal region, as it approaches the temporal neocortex (Fig. 9.1g–i). It is noteworthy that the two most heavily involved layers (pre-α and pri-α) are chiefly responsible for the bidirectional transfer of data between the neocortex, entorhinal region, and hippocampal formation (Fig. 6.4c, d).

The presubiculum and subiculum are still uninvolved. At the same time, the involvement of pyramidal cells in the superficial layer of CA 1 and in CA 2 becomes more pronounced: Transient pathological alterations develop in apical dendrites of the outer CA 1 pyramidal cells. Conspicuous spindle-shaped dilations develop in the *stratum lacunosum-moleculare* and are filled, initially, with AT8-ir inclusions and, later, argyrophilic material (Fig. 9.5). These dilations first appear in a few cells during NFT stage II (Fig. 9.5a) and are most pronounced in stage III (Fig. 9.5b) (Braak and Braak 1997a). Such varicose segments vanish from the tissue during NFT stages IV and V without leaving behind any remnants, probably because the dendritic material can be eliminated (as opposed to tombstone tangles). The end result is that the contact zone between the perforant pathway and CA 1 pyramidal cells is lost, thereby partially disconnecting the neocortex from the hippocampal formation (Kemper 1984; Hyman et al. 1988, 1990; van Hoesen and Hyman 1990; Delbeuck et al. 2003).

The CA 2 sector usually becomes filled with strongly AT8-ir pyramidal cells during stage III. Frequently, but not always, CA 2 is the earliest sector involved. The appearance and development of the tau lesions in CA 3/CA 4 lags behind those in CA 1/CA 2. The first AT8-ir mossy cells appear in CA 3/CA 4 with star-shaped NFTs that fill the characteristically large dendritic excrescences (Blazquez-Llorca et al. 2011). In the dentate fascia, the granule cells remain uninvolved.

From the transentorhinal region, the lesions encroach upon temporal neocortical areas adjoining the transentorhinal region laterally (Fig. 9.2a, b) and posteriorly, i.e., in areas covering the fusiform and lingual gyri (Fig. 9.2b). Neocortical involvement during stage III is confined to the basal regions of the temporal lobe and diminishes markedly beyond them (Fig. 9.2b). The propagation of the disease process into the mature neocortex probably takes place via cortico-cortical projections of the return pathway and is supported by diffuse projections from subcortical sites (Fig. 6.8e). Consistent with the fact that cortical layer IV is virtually spared in AD, the spreading of the pathology from periallocortical fields into the temporal proneocortex and neocortex is likely to occur via the pyramidal cells in layer V, whose axons mostly terminate in layers I, III, and VI of their target areas and have very few interconnectivities with layer IV (Fig. 6.8e).

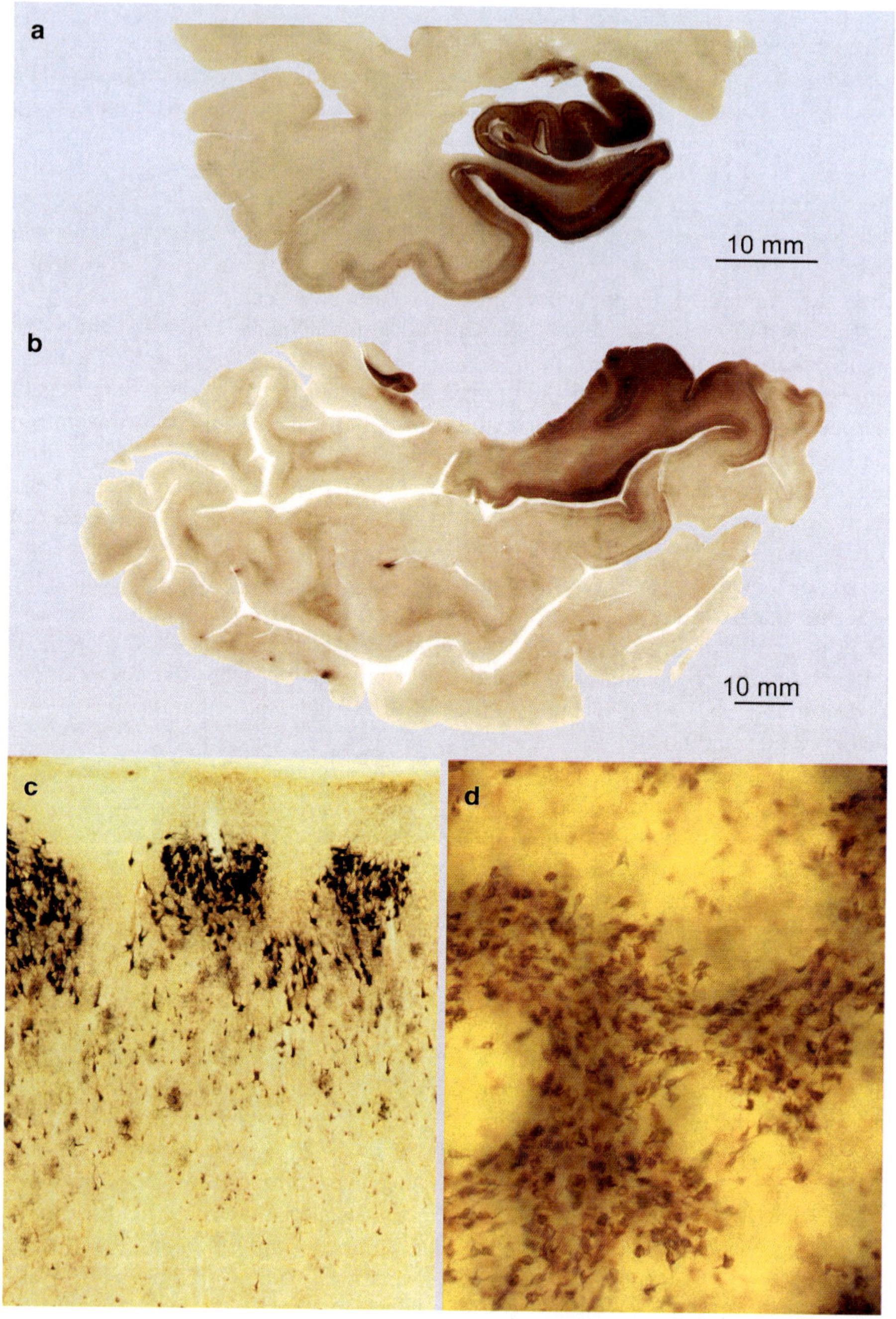

Fig. 9.2 Overview of the extent of tau pathology during NFT stage III in 100 μm sections. (**a**) The hippocampal formation, entorhinal region, and transentorhinal regions are heavily involved in cases showing NFT stage III. From there, the lesions characteristically progress into the adjoining basal temporal neocortex that covers the occipitotemporal gyrus. Note that the density of the pathology gradually decreases and generally does not reach the inferior temporal gyrus (65-year-

9.2 Involvement of Neocortical Chandelier Cells

During NFT stage III, while the pathological process progresses from the transentorhinal region into adjacent portions of the basal temporal neocortex, chandelier cells in these involved areas also develop abnormal tau (Braak et al. 2006b). Elsewhere, such local circuit neurons are not known to become involved in the AD process (see Sect. 2.2). Nonetheless, their highly atypical reaction at precisely this site and at this stage is noteworthy and possibly of importance for the pathogenesis of AD.

Inhibitory cortical chandelier cells belong to the class of gabaergic local circuit neurons with smooth dendrites, i.e., dendrites that do not have spiny appendages. Chandelier cells usually remain devoid of lipofuscin granules or develop only a few (Braak 1980). Their distinctively arborized axons (Fig. 9.3a, arrow) give off vertically-oriented rows of symmetrical synapses attached to the initial segment of axons of cortical pyramidal cells. By means of these axo-axonic contacts, the chandelier cells can inhibit the generation of action potentials (Howard et al. 2005; Woodruff et al. 2010; Benarroch 2013). Among the various forms of local circuit neurons and other neuronal types that have few or no lipofuscin granules, the chandelier cells are the only ones in which non-fibrillar AT8-ir material occasionally develops (Braak et al. 2006b).

Involved chandelier cells usually are seen in the superficial layers (II–III) of the basal temporal lobe, where the AT8-ir material, notably, initially appears in and mostly remains confined to the axon. Nevertheless, the typical axonal morphology permits a reliable diagnosis of the chandelier cell type (Fig. 9.3b–e). On occasion, the AT8-ir axon with all its vertically-oriented terminal portions is accompanied by AT8-ir material in the soma, and a few traces may even appear in dendritic processes (Fig. 9.3c). It is important to note that, in this type of local circuit neuron, the non-fibrillar tau material does not convert into argyrophilic inclusions. AT8-ir chandelier cells probably perish rather quickly because they are not seen any more after NFT stage IV.

What clues do the chandelier cells reveal with regard to the pathological process? First, it is probably only the chandelier cells that synapse with axons of involved cortical pyramidal cells that become involved. Here, too, it can be speculated that abnormal tau inclusions in chandelier cells may be actively induced via the axon of affected pyramidal cells: Pathogenic tau material may pass through

Fig. 9.2 (continued) old male, NFT stage III, Aβ phase 2). (**b**) This tangential section shows superficial portions of the entire temporal lobe. Most densely affected is the entorhinal region of the parahippocampal gyrus, from where lesions progress to a limited extent into adjoining fields of the basal temporal neocortex (89-year-old female, NFT stage III, Aβ phase 1). (**c**) In NFT stage III cases, the cellular islands of the pre-α layer are particularly heavily involved and may even show the first tombstone tangles. The subjacent layer pre-β begins to show pathological changes. (**d**) Flat section providing a bird's eye perspective of layer pre-α with its leopard skin-like pattern. AT8 immunoreactions (**a**, **b**) and Gallyas silver-iodide technique (**c**, **d**)

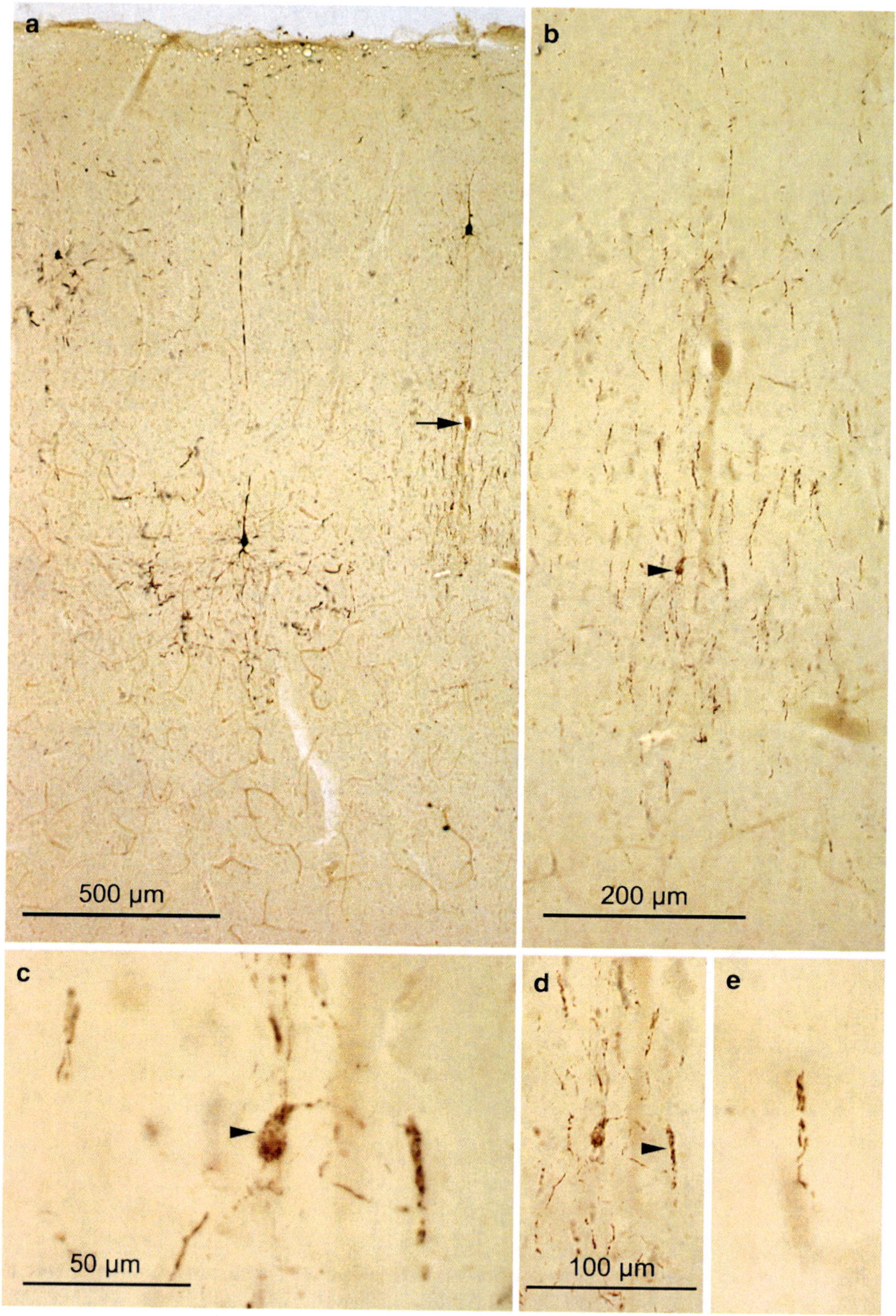

Fig. 9.3 Chandelier cells in 100 µm AT8-immunostained sections. (**a**) Overview of the basal temporal neocortex covering the occipito-temporal gyrus (54-year-old female). A few AT8-ir pyramidal cells in layer III display typical early reactions in their distal dendritic segments. *Arrow* points to a nearby ensemble of axon terminals belonging to an AT8-ir chandelier cell. (**b**) The same cell in greater detail. A large number of candle-like axon terminals is loosely arranged

the membranes of the axonal initial segment into the terminal axon of chandelier cells. This would also explain why the AT8-ir material that develops in chandelier cells initially occurs only in the axon and appears only in exceptional instances within the cell soma and dendrites. Second, small amounts of non-fibrillar abnormal tau in certain neurons apparently can result in cell death because abnormal chandelier cells are no longer seen at later disease stages. Moreover, their somatodendritic compartment is incapable of converting abnormal tau into argyrophilic filaments. Such cells may die prematurely because a putative 'detoxification' of the non-fibrillar material via a conversion into fibrillar inclusions fails to take place.

Chandelier cells are exceptional because they establish and maintain an especially close contact to the axons of cortical pyramidal cells. Thus, chandelier cells could be helpful in resolving certain questions pertaining to the various effects of aggregated tau proteins on diverse types of nerve cells. It may well be that each and every type of nerve cell reacts differently to the sudden presence of potentially harmful hyperphosphorylated oligomeric tau.

9.3 Are Stages a–III Part of the AD-Associated Pathological Process?

From stage I onwards, all of the involved predilection sites in the CNS consistently display the following neuronal combination: nerve cells that resist the AD process, uninvolved but potentially susceptible neurons, neurons that contain non-argyrophilic AT8-ir pretangle material, argyrophilic NFT-bearing neurons, and their tombstone remnants. Were the AD process to be characterized by longer periods of remission or spontaneous healing, two morphological additional traits should be present, regardless which NFT stage has been reached. First, the pathological process in all involved neurons should consist only of argyrophilic NTs/NFTs and tombstone tangles, and, second, uninvolved (but potentially susceptible) nerve cells should not generate fresh lesions consisting of non-argyrophilic pretangle material. NTs/NFTs in the absence of pretangle material would indicate that the pathological process was active in the brain for an indefinite period of time, but that formation of new tangles had ceased. However, it must be emphasized that this situation does not occur. This fact bears out our interpretation that the pathological process persists from its outset until death and, in so doing, virtually eliminates the possibility of spontaneous remission. Cases with dissimilar NFT stages in double-hemisphere sections are also instructive in this regard. Figure 2.4c, for example, shows one of the hemispheres with lesions corresponding to NFT

Fig. 9.3 (continued) around the cell body (*arrowhead*). (**c**) The soma (*arrowhead*) gives rise to thin AT8-ir dendrites. (**d**) Detail of the single candle-like formation (*arrowhead*) seen in detail in (**e**)

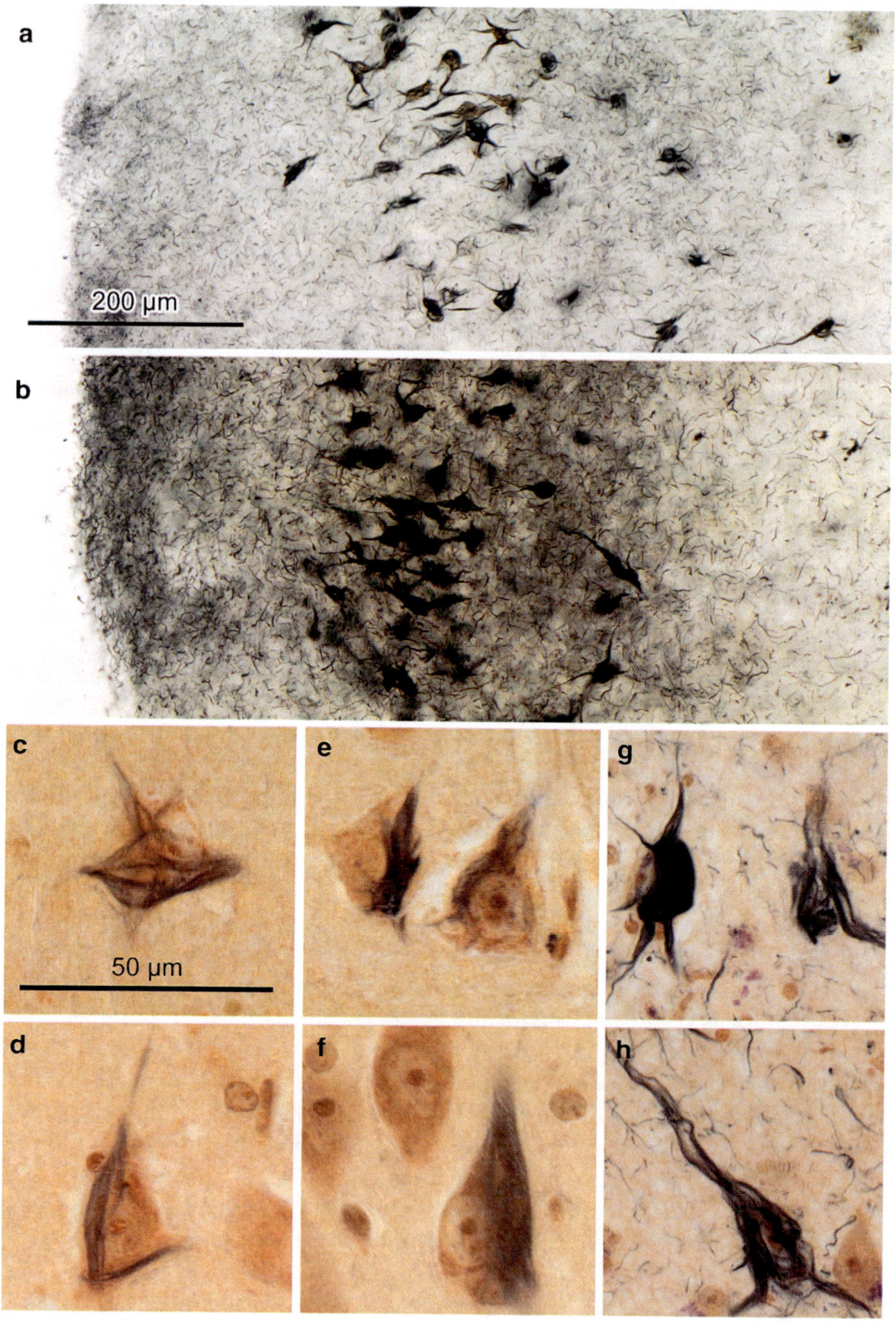

Fig. 9.4 Neurofibrillary pathology in the entorhinal layer pre-α as seen in 100 μm sections (Gallyas silver-iodide technique). (**a**) Mildly involved pre-α at NFT stage II. Mature NFTs are seen in projection cells of layer pre-α, and a dense network of NTs begins to develop in superficial portions of the layer and the adjacent molecular layer (NFT stage II). (**b**) A dense network of NTs

stage II and the other to NFT stage IV (see also Fig. 2.4b). It would be absurd to assume that only one of the hemispheres (namely, that at stage IV) is involved in the AD process, whereas the stage II lesions represent a non-AD-related variant of aging. Indeed, such cases provide evidence within one and the same individual for the existence of different stages of a single pathological process during its gradual but continual development.

It has long been discussed whether early tau stages without Aβ deposits ('NFT-only' or 'tau-only' cases) and, more recently, stages a–1b, belong to the process with the potential to cause clinical symptoms of AD (Price et al. 1991; Dickson 1997a; Price and Morris 1999; Jack et al. 2013). The conceptual basis for excluding such cases from the AD process is the assumption that Aβ 'drives' tau aggregation, so that the threshold to clinical AD is crossed only in the presence of Aβ deposits (Price and Morris 1999, 2004; Sperling et al. 2011; Jack et al. 2013). As such, it is reasoned that tau aggregates either in the absence of Aβ or as long as they remain below the detection threshold for available biomarkers of MCI/AD, represent either a 'benign variant' of normal aging or a non-AD-related tauopathy (Sperling et al. 2011; Jack et al. 2013; Mann and Hardy 2013; Thal et al. 2013; Crary et al. 2014) rather than potential early stages of a neuropathological continuum (Schönheit et al. 2004; Braak and Del Tredici 2011, 2013b, 2014; Braak et al. 2011; Duyckaerts 2011).

In fact, a surprisingly large number of non-cognitively impaired individuals have mild tau lesions in the absence of Aβ plaques and NPs and, remarkably, such cases are found in all age categories (Tables 7.2 and 9.4; see also Sect. 8.6). In order to classify these cases as non-AD-related, evidence is still required proving that the tau pathology in such individuals *ceases to develop beyond* NFT stages III or IV. However, cases with NTs/NFTs, tombstone tangles, and glial scares persisting in the absence of non-argyrophilic AT8-ir pretangle lesions have not been described to date, nor do they occur in the cohort here. As such, their existence is quite unlikely. Even more unlikely is the assumption that, as soon as Aβ deposition begins in phase 1 in the presence of tau pathology corresponding to NFT stages II–IV, enormous numbers of NTs/NFTS (including even tombstone tangles), could immediately develop based on the existence of a few plaques in the basal temporal and frontal cortex. In addition, during the transition from phase 1 to phase 2 Aβ deposition, one would expect to see a continuation of the initially fulminant

Fig. 9.4 (continued) is seen above heavily involved layer pre-α cells at NFT stage V. The network of NTs loses both its immunoreactivity as well as its argyrophilia because it consists to a large extent of dendritic branches that no longer are connected to their proximal stems, i.e., tombstone NTs (82-year-old female AD patient, NFT stage V). (**c–h**) Development of neurofibrillary changes in layer pre-α as seen using the Gallyas silver-iodide technique combined with pigment-Nissl staining (aldehydefuchsin and *Darrow red*). (**c**, **d**) Initially, NFTs in pre-α cells cause no clearly negative response (60-year-old male, NFT stage I). (**e**, **f**) Consolidation of a sturdy NFT. A comparison with uninvolved neighboring neurons does not reveal obvious reactive changes (same individual as in **c** and **d**). (**g**, **h**) Dying NFT-bearing nerve cell at the *left side* in (**g**) allows comparison with tombstone tangles in (**g**) and (**h**) (60-year-old male AD patient, NFT stage VI). Scale bar in (**a**) applies to (**b**). Scale bar in (**c**) applies to micrographs (**d–h**)

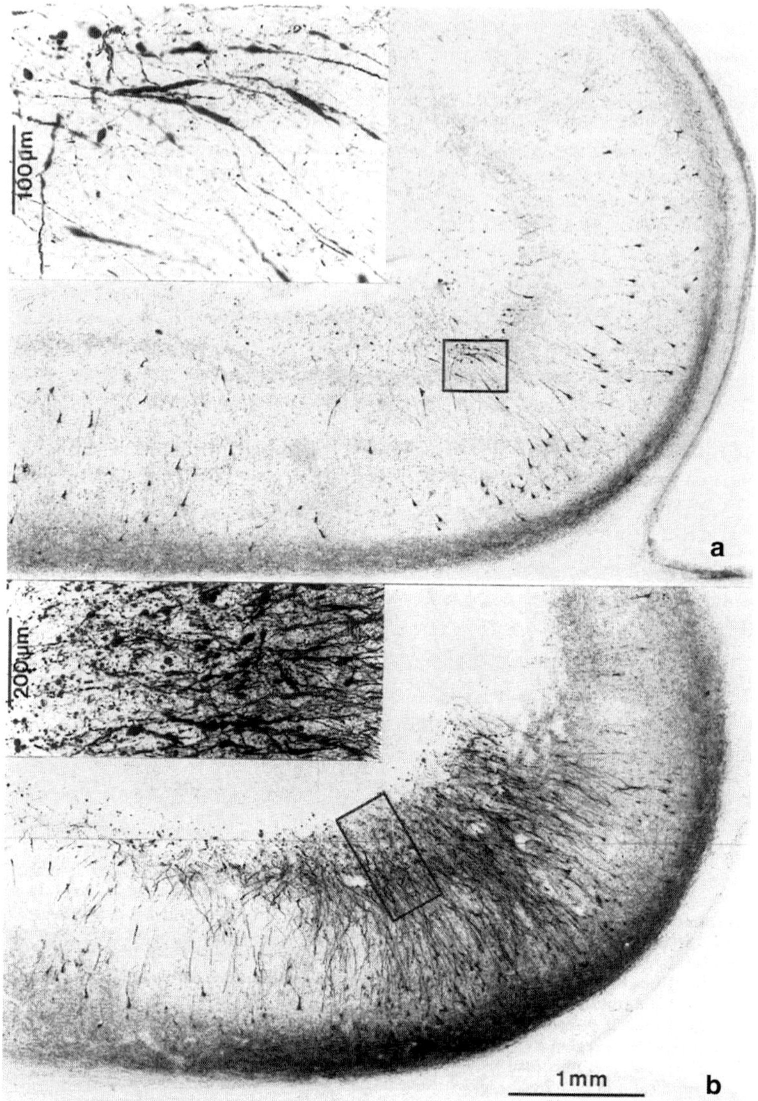

Fig. 9.5 CA1 pyramidal cells with transient dendritic spindles in AT8-immunostained 80 μm sections. The apical dendrites of CA1 pyramidal cells develop *spindle*-shaped dilations (the insets correspond to the framed areas at higher magnification). (**a**) The *spindle*-shaped dilations appear in NFT stage II and are most developed in NFT stage III in (**b**) (reproduced with permission from H Braak and E Braak, Temporal sequence of Alzheimer's disease-related pathology. Cerebral Cortex 1999;14:475–512)

development of the tau lesions (i.e., because Aβ deposition is present to a larger extent). But, this too, does not occur.

Here, 1,253 of the 1,885 cases corresponding to stages a-II are devoid of Aβ plaque deposition (=approximately 66 %) and these increase to 1,509 cases

(=approximately 80 %) when cases with sparse plaques (phase 1) are included (Table 7.2). The tau aggregates in such cases consist of both 3R and 4R isoforms (Iseki et al. 2002; Jellinger and Attems 2007) and, thus, are morphologically indistinguishable from those evolving in the presence of Aβ deposits or in more advanced stages of the AD process. Moreover, they only occur at known AD predilection sites and they only develop in neuronal types known to be vulnerable to the AD process. In other words, the mere absence of Aβ deposition is not an adequate rationale for excluding tau-only cases from the developmental spectrum of AD-associated preclinical stages (Table 7.2), nor is the presence of such cases at younger age categories consistent with the argument that tau aggregates in tau-only cases are a benign variant of normal aging (Davis et al. 1999; Jack et al. 2013; Korczyn 2013). The existence of a single malignant cell, no matter at what age or what its prevalence in a given population, does not make it less malignant.

Equally unconvincing is the argument that tau-only cases might represent preclinical forms of rare non-AD tauopathies, such as PiD, PSP, or CBD (Mann and Hardy 2013; Thal et al. 2013). Alone the large number of tau-only cases (Table 7.2) and the presence of both 3R and 4R isoforms as opposed to that of the 3R (PiD) or 4R isoform (PSP, CBD) speaks against the assumption that they are possible manifestations of non-AD tauopathies (Uchihara et al. 2005, 2011, 2012; Braak and Del Tredici 2013b). No staging procedures have been proposed or validated for these disorders as possible frames of reference, thereby rendering a definition of putative 'preclinical' PiD, PSP, and CBD impossible as long as firmer knowledge of what may define their 'early' pathologies is lacking. Assuming, however, that such prodromal stages do exist, we expect that their hallmark lesions will develop in susceptible cell types and at regional predilection sites that are characteristic of PiD, PSP, and CBD: in preclinical PiD, for example, Gallyas-negative tau aggregates in granule cells of the dentate fascia and, in PSP, Gallyas-positive tau inclusions in astrocytes (Dickson et al. 2007). However, in contrast to these disorders, the features of tau-only cases shown in Table 7.2 have strong ties to the AD process. Finally, evidence also exists for a strong link between tau-only cases and the APOE ε4 allele: In a previous study, cases with NFT stage I pathology displayed a significantly higher APOE ε4 allele frequency than controls (Ghebremedhin et al. 1998). There is no cogent reason, therefore, why stages a-III should be excluded from the natural history of the AD-related pathological process (Braak and Del Tredici 2014).

9.4 Basic Organization of Insular, Subgenual, and Anterior Cingulate Regions

The insular, subgenual, and anterior cingulate regions represent the highest organizational level of the cerebral cortex processing interoceptive data and regulating visceromotor as well as endocrine functions. These regions are characterized by conspicuously slender and spindle-shaped von Economo neurons in layer V. They evolve in primates preferably in the hominoid lineage and probably are involved in autonomic regulation (Butti et al. 2013).

The agranular and dysgranular regions of the insula, together with the adjoining association areas, encompass gustatory areas and a topically-organized representation of the internal organs and inner surface of the body. They are reciprocally connected with subgenual, anterogenual, and anterior cingulate areas, the entorhinal region, amygdala, claustrum, thalamic limitans nucleus, and the pigmented parabrachial nucleus. They also generate major projections to the magnocellular nuclei of the basal forebrain and ventral striatum (Fig. 6.10a). In this manner, a pathway is established between the insular fields – via ventral striatum, ventral pallidum, and mediodorsal thalamus – to the prefrontal association cortex. Via the claustrum and magnocellular nuclei of the basal forebrain, the insular fields also exert their influence on the cerebral cortex as a whole. For this reason, the agranular and dysgranular regions of the insula bear the designation 'viscerosensory and limbic integration cortex' (Mesulam and Mufson 1993; Nieuwenhuys 2012).

The subgenual, anterogenual, and anterior cingulate regions are part of the medial frontal lobe and represent topically organized visceromotor centers. Bidirectionally organized projections connect the regions with adjoining prefrontal areas, the insular cortex, entorhinal region, hippocampal formation, amygdala, intralaminar and midline nuclei of the thalamus, lateral hypothalamus, periaqueductal nuclei, and autonomic regions of the lower brainstem and spinal cord. These regions again send strong projections to the ventral striatum and, consequently, act upon the prefrontal cortex via ventral pallidum and mediodorsal thalamus. As such, the regions are regarded as fulfilling the functions of a 'visceromotor and limbic integration cortex' (Vogt 2009; Vogt et al. 1993; Price et al. 1996).

9.5 NFT Stage IV: Further Progression of the Lesions into Proneocortical and Neocortical Regions Governing High Order Autonomic Functions

During stage IV, the tau pathology extends more widely into the proneocortex and neocortex. From the previously involved basal temporal fields, the disease process encroaches upon insular, subgenual, anterogenual, and anterior cingulate areas (Vogt 2009). The neurofibrillary lesions in the dysgranular insular fields are less severe than those in the agranular portions (Bonthius et al. 2005). Lesions in temporal areas are most dense up to the middle temporal convolution but rapidly decrease at the transition from the sparsely myelinated middle to the highly-myelinated superior temporal gyrus (Fig. 9.13, stage IV). The occipital neocortex is still largely uninvolved or displays localized blotch-like accumulations of AT8-ir pyramidal cells and/or NPs in high order sensory association fields.

The density of AT8-ir feltworks in the entorhinal and transentorhinal regions increases, which leads to a blurring of the *lamina dissecans* (Fig. 9.1k–m). The severity of the lesions in these regions and also within the hippocampal formation peaks during this stage, but apart from a slight increase of pathology, the major features of the allocortical pathology remain essentially the same from stage IV onwards. Among the most noticeable developments is the increasing involvement of the external layers pre-β and pre-γ as well as of the deep layer pri-γ. At this point, numerous large NPs begin to appear in pre-β and pre-γ, while the superficial layer pre-α is spared.

The density of the deep plexus of AT8-ir cellular processes spanning all of the deep layers pri-α, pri-β, and pri-γ reaches its culmination point. Nerve cells in the white matter (*lamina cellularis profunda*) also become involved, although they do not attract the viewer's attention immediately because their density is appreciably lower than that of cells in the deep layers.

In stage IV, the small projection neurons of the parasubicular portion of the presubiculum develop tau aggregates. A few AT8-ir pyramidal cells appear for the first time in the subiculum (Fig. 9.1k–m). The CA 1/CA 2 sectors are filled with NTs/NFTs and are recognizable as dense bands or stripes. The first tombstone tangles also can be seen in the superficial layer of CA 1. The varicose dendritic segments vanish from CA 1 without leaving behind any remnants. Portions of the neuropil that normally contain axons of the perforant path appear very fragile, and tears are frequently visible there in tissue sections. Surviving pyramidal neurons in CA 2 and those in adjacent portions of CA 1 display signs of granulovacuolar degeneration (Thal et al. 2013). Large numbers of mossy cells in CA 3 and CA 4 develop neurofibrillary lesions. A few AT8-ir granule cells appear for the first time in the dentate fascia.

The subcortical lesions in the nuclei with diffuse cortical projections are severe at this stage and, to a great extent, they already consist of argyrophilic NTs/NFTs. The damage within the limbic nuclei of the thalamus (reuniens nucleus, anterodorsal nucleus, limitans nucleus) intensifies and supplements that in the superordinate nuclei responsible for regulating autonomic function (e.g., central nucleus of the amygdala, the parabrachial nuclei) (Parvizi et al. 1998; Blessing 2004). Moreover, one regularly encounters involvement of the large cholinergic interneurons within the putamen as well as the caudate and accumbens nuclei. The presence of tau aggregates makes the dendritic trees of these cells remarkably prominent. The neurons of the pallidum remain univolved and, in the amygdala, pathology begins to accrue in the basolateral complex of the amygdala. The lesions in the central and cortical subnuclei become worse, and only the medial subnuclei of the amgydala are still nearly intact. In the tectum, numerous NPs develop, above all in the inferior colliculus. Isolated AT8-ir melanized neurons are visible in the pars compacta of the substantia nigra (Fig. 9.6h). A moderate number of α-motoneurons in the spinal cord are accompanied by an increasing network of AT8-ir neuronal processes (probably descending axons lower raphe nuclei and the subcoeruleus nucleus) (Fig. 9.6b, c).

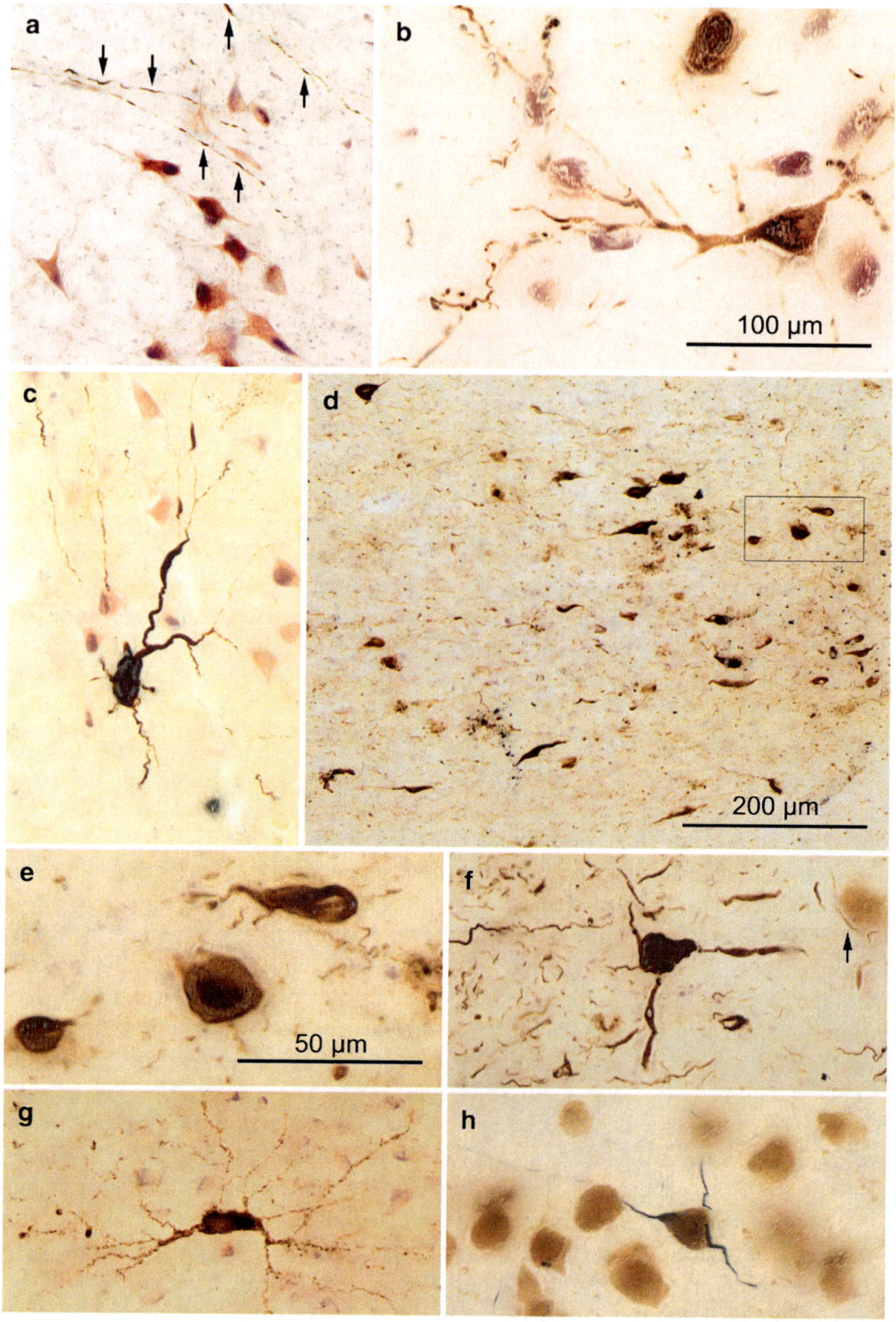

Fig. 9.6 AD-related tau pathology in the spinal cord, olfactory bulb, and substantia nigra. (**a**) *Arrows* indicate a faintly AT8-ir neuritic network, probably consisting of axons, in the spinal cord gray matter (segment Th1) of a 74-year-old male (NFT stage II). (**b**) Two involved AT8-ir α-motoneurons, including an AT8-ir somatodendritic compartment, in segment C3 of a 79-year-old female at NFT stage III. In (**c**), a severely involved α-motoneuron is seen in the ventral horn

9.6 Macroscopically Recognizable Characteristics of Advanced AD

The first indices of a possibly atrophic process are visible upon macroscopic examination of the brain (compare control, Fig. 9.7a with AD brain, Fig. 9.7b). The signs are bilaterally symmetrically remarkable and are seen especially in anteromedial regions of the frontal and temporal lobes, particularly in the entorhinal region of the parahippocampal gyrus. The *verrucae hippocampi* (Fig. 6.4a) are flattened and barely recognizable (Simic et al. 2005). By contrast, atrophic changes cannot be recognized in pre- and postcentral gyri, the transverse gyri of Heschl, and in the vicinity of the calcarine sulcus. Moreover, the ventricular system is widened, particularly the temporal horn (Figs. 2.3c and 9.14). Widening of the ventricular system with thinning of the hippocampal formation can also be visualized using MRI scans (Dickerson et al. 2011). In the event that the clinical records document the existence of a dementing process, such signs can provide an initial indication that an AD-associated pathological process could be present.

9.7 NFT Stage V: Fan-Like Progression of the Neocortical Pathology into Frontal, Superolateral, and Occipital Directions and its Encroachment on Prefrontal and High Order Sensory Association Areas

The main feature of stage V is severe involvement of large portions of the neocortex, leaving only the primary sensory and motor fields and their belt regions uninvolved or mildly affected (Fig. 9.14). From neocortical sites already involved at stage IV (regions for high order cortical autonomic regulation), the lesions extend widely into prefrontal fields and high order sensory association areas of the temporal, parietal, and occipital neocortex. The occipital lobe presents a well preserved primary visual area on both banks of the calcarine sulcus (Fig. 9.14 upper right), followed by a mildly involved parastriate field (border field), and a heavily affected peristriate region (high order association fields).

Fig. 9.6 (continued) (segment L1) of a 77-year-old male (NFT stage II) with concomittant Parkinson's disease (PD stage 5). (**d–f**) Extensive tau pathology in the olfactory bulb of a 76-year-old male (NFT stage III), including the anterior olfactory nucleus (**d**) and large projection neurons (framed area in **d** can be seen at higher magnification in **e**). (**f**) Note the heavily involved mitral cell adjacent to a completely normal one (*arrow*). The olfactory glomerula and granule cells appear to remain unaffected. (**g**) Detail micrograph of a AT8-ir cell in the anterior olfactory nucleus of a 59-year-old male at tau stage 1b. (**h**) Late AD stages consistently show tau pathology in dopaminergic projection neurons of the substantia nigra, pars compacta (74-year-old male, NFT stage IV) 100 μm sections. AT8-immunoreactions combined with pigment-Nissl staining. Scale bar in (**d**) applies to (**a**) and (**c**), and scale bar in (**b**) applies to (**f**), (**g**), and (**h**). Scale bar in (**e**) also applies to (**f–g**)

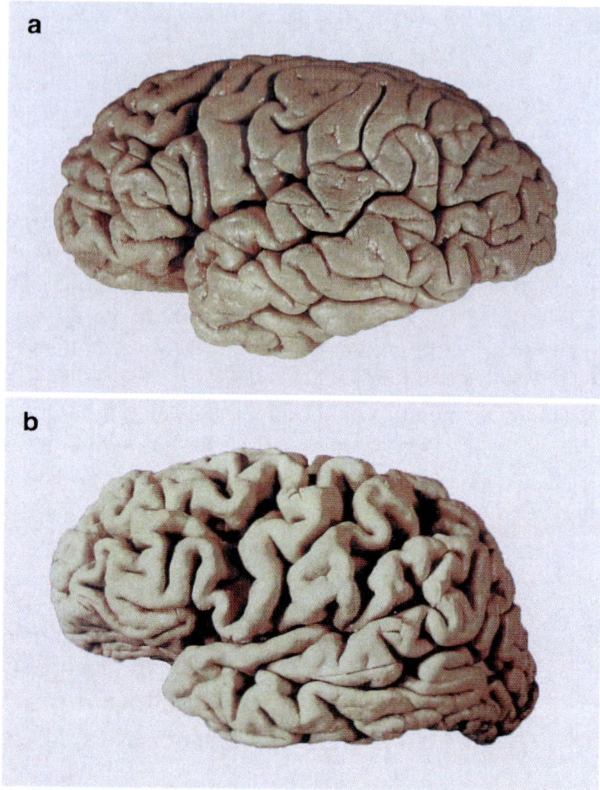

Fig. 9.7 Comparison of a brain of a non-demented individual with that of an AD patient. The AD-brain shows remarkable widening of the sulci and narrowing of the gyri in all lobes. Note the extensive and severe atrophy of the temporal lobe

Initially, unevenly and loosely distributed NPs appear in layers II and III of the peristriate areas, followed by large numbers of AT8-ir pyramidal cells in layers IIIa, b and V, for the most part already containing argyrophilic NTs/NFTs. The lower border of the outer neuritic plexus in layer IIIa,b blurs at its transition to the less heavily involved layers IIIc and IV (outer line of Baillarger). In stage V, the deep plexus of layer Vb (inner line of Baillarger) is narrow and does not tend to extend into layer VI and the white matter. The same pattern (only less pronounced) is seen in the parastriate border field, where uneven accumulations of NPs predominate.

9.8 NFT Stage VI: The Pathological Process Progresses Through Premotor and First Order Sensory Association Areas into the Primary Fields of the Neocortex

The pathological process reaches its greatest extent during the end-stage VI (Figs. 2.3c and 9.14). Deviations from a bilaterally symmetrical distribution of the pathology do not occur during stage VI. The intact or only mildly involved premotor and first order sensory association areas of the neocortex clearly show nerve cell destruction, and the pathological process advances from these areas into the primary fields (see Fig. 9.14, lower right: striate area 17, and Fig. 9.9).

The neocortex is severely involved: Pyramidal cells in layers III, Va, and VI are the hardest hit cell types. Nearly all cellular layers are filled with AT8-ir neuronal processes. As such, even the outer line of Baillarger – still a pallid stripe in stage V – begins to blur (compare peristriate fields area 19 in stage V at upper right with those in stage VI at lower right, large red arrows). Layer Vb appears as a recognizable band but continues into the neuritic plexus of layer VI. The underlying white substance contains abundant AT8-ir axons. NPs display decreased immunoreactivity and Gallyas-positive argyrophilia in many neocortical areas. This latter feature, which is most pronounced in the basal temporal field, probably indicates their degradation.

In the occipital lobe, the pathology breaches the parastriate region (Brodmann area 18) and the striate area (Brodmann field 17) (Braak et al. 1989a). The line of Gennari in the striate area maintains a light appearance (Figs. 9.14, lower right and 9.9d), interrupted only by radially oriented AT8-ir axons (Fig. 9.9e). A sharply delineated AT8-ir plexus, usually a dense network of NTs, follows in the narrow layer V (Fig. 9.9d, e). This layer displays very severe Aβ deposition (Fig. 9.9a) and NPs (Fig. 9.9d), whereas the line of Gennari harbors only Aβ plaques and dot-like deposits of Aβ (Fig. 9.9b, c) that tend not to converge into Gallyas-positive NPs (compare Aβ in Fig. 9.9a–c with tau in Fig. 9.9d, e). The density of NTs/NFTs increases little by little proceeding from the primary visual field via the parastriate region into occipital high order sensory association areas. The boundary between the parastriate field and the peristriate region, which is not easy to identify in the normal human brain, is light microscopically recognizable in end-stage AD cases (Braak et al. 1989a).

The situation in the allocortex essentially reveals an increase of the pathology seen there previously in stage V (Fig. 9.8). In the entorhinal region, layer pre-α occasionally appears denuded of nerve cells, with tombstone tangles being the only remnants (Fig. 9.4g, h). An additional feature that distinguishes NFT stage VI from stage V is the large number of argyrophilic globose NFTs in granule cells of the dentate fascia.

The colliculi of the midbrain show large numbers of tau lesions, Aβ plaques, and NPs (Dugger et al. 2011). A sizeable proportion of neuromelanin-laden neurons in the pars compacta of the substantia nigra display NTs/NFTs (Fig. 9.6h).

9.9 The Pattern of the Cortical Tau Pathology in AD Mimics the Developmental Sequence of Cortical Lipofuscin Deposits and, in Reverse Order, That of Cortical Myelination

The sequence of AD-lesions follows a pattern of progression (Fig. 9.8a) similar to that seen in the appearance of lipofuscin deposits in cortical projection neurons (Fig. 6.8c), whereas the evolution of cortical myelin (Fig. 6.8a) reiterates this

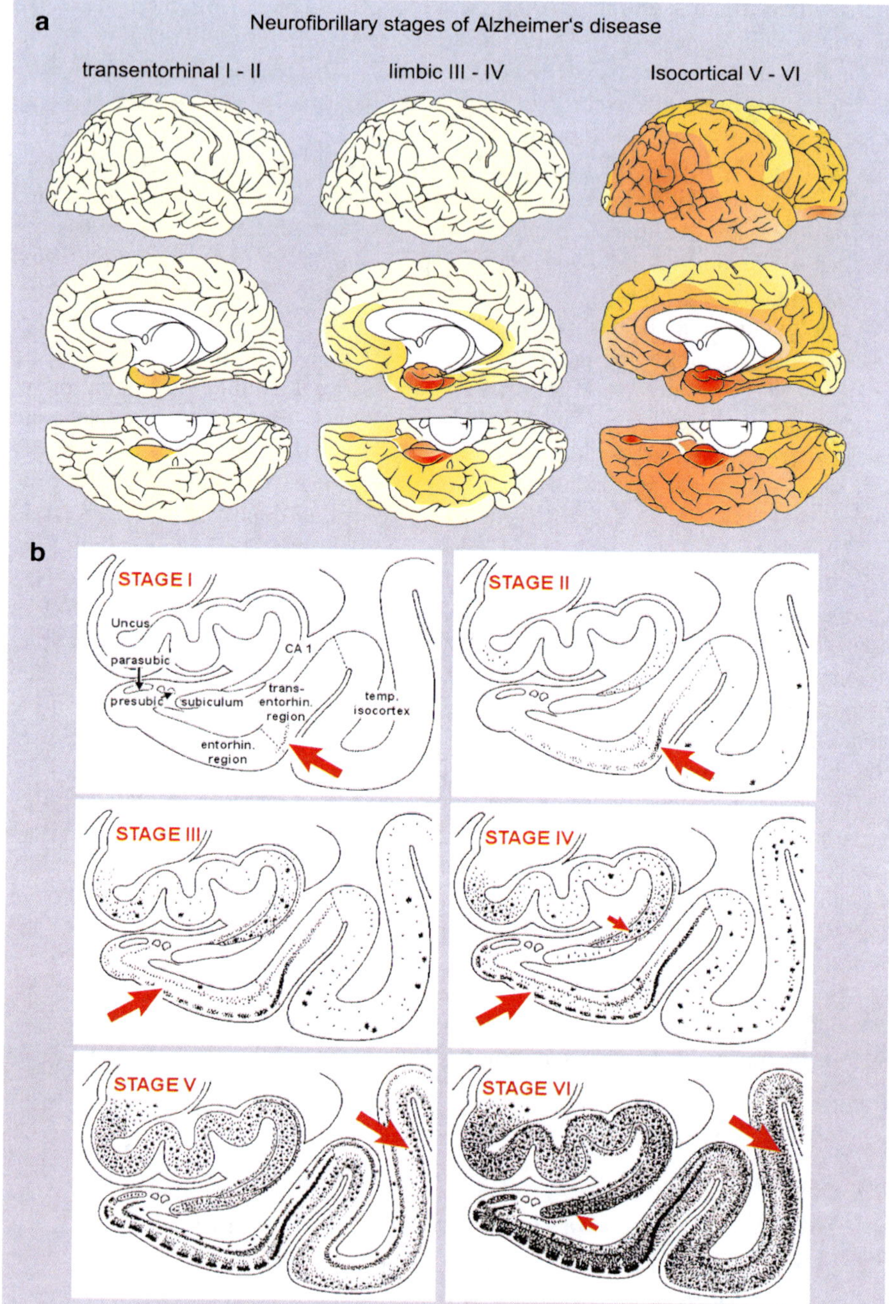

Fig. 9.8 Distribution pattern of AD-related argyrophilic intraneuronal lesions. (**a**) Six NFT stages can be distinguished. NFT stages I–II show abnormal alterations that are virtually confined to a single layer of the transentorhinal and entorhinal regions. NFT stages III and IV display additional involvement of many cortical regions and subcortical nuclei of the limbic system. Stages V–VI are

sequence: The progression is the same, but the order is reversed (Fig. 6.8b). The primary neocortical fields myelinate first and, thus, in the human adult they are especially heavily myelinated. Myelin density gradually declines via the belt regions into the high order association areas, i.e., declines with increasing distance from the primary areas (see Sect. 2.2). Just as the heavily myelinated but sparsely pigmented primary neocortical fields are more or less impervious to the AD process, cortical areas rich in pigment and that myelinate last are the most prone to develop the tau lesions (Fig. 6.8b). As such, it is not surprising that the poorly myelinated and heavily pigmented anteromedial and basal temporal areas, including the transentorhinal region, are the sites where the earliest cortical AD-lesions develop (Fig. 6.8b). From there, the pathology slowly progresses and extends into hitherto uninvolved portions of the cortex in the opposite direction of the myelination process (Braak and Braak 1996). This developmental pattern supports the observation that regressive brain changes tend to repeat the maturation process in reverse order (retrogenesis), and, quite remarkably, the pattern in the CNS also is reflected in the antidromic development (i.e., diminution) of individual life skills during the course of the AD-associated dementive process (Table 9.1) (Reisberg et al. 1999, 2002; Arendt et al. 1998; Cramer and Chopp 2000; Moceri et al. 2000; Braak and Del Tredici 2004; Stricker et al. 2009; Gogtay and Thompson 2010; Ewers et al. 2011; Ashford and Bayley 2013; Rubial-Álvarez et al. 2013).

9.10 The Prevalence of Tau Stages and Aβ Phases in Various Age Categories and Potential Functional Consequences of the Lesions

Taken together, the following two figures summarize the proposed sequence of the AD pathological process from stage a to stage VI (Figs. 9.10 and 9.11). Table 9.2 contains data pertaining to the frequency of all tau stages separated by gender (Braak and Braak 1997b; Duyckaerts and Hauw 1997; Dickson 1997a; Hyman and Gómez-Isla 1997).

Cases with lesions corresponding to stages a–1b and I–II represent a clinically silent period of the AD process (Figs. 2.1d, 9.10a, b, 9.13, and 9.15a, b). The degree of damage caused by these early neurofibrillary lesions cannot be determined or quantified by currently available diagnostic tests and instruments. Thus, what is needed is the discovery of practical methods for ascertaining the possible existence

Fig. 9.8 (continued) marked by devastating destruction of the neocortex. (**b**) Summary diagram showing the development of intraneuronal tau lesions from NFT stage I–VI in the hippocampal formation, entorhinal and transentorhinal regions, as well as in the adjoining temporal neocortex. *Arrows* point to diagnostically distinguishing marks of the pathology for each stage. Abbreviations: *CA 1* first sector of the Ammon's horn, *parasub* parasubiculum, *presubic* presubiculum, *transentorhin* transentorhinal, *entorhin* entorhinal, *temp* temporal neocortex

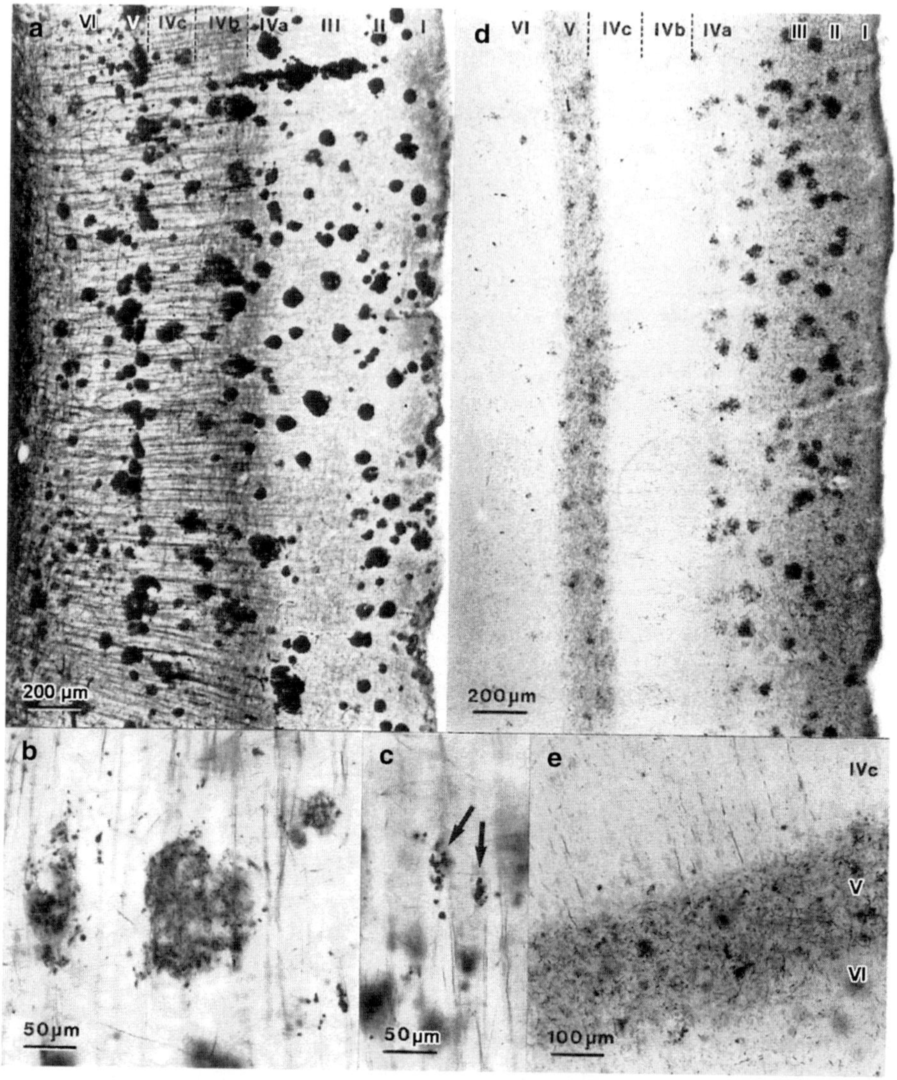

Fig. 9.9 Striate area in silver-stained 100 μm sections. Note that the laminar distribution of Aβ deposits (**a**–**c**) differs considerably from that of neurofibrillary lesions (**d**–**e**). Layers are indicated at the upper margin in Roman numerals. (**b**, **c**) Weakly stained Aβ plaques and intensely marked dots (*arrows*) in layer IVc that do not convert into NPs. (**e**) Dense line of NTs characterizes layer V. Note the vertical NTs extending from this line into portions of layer IVc. Campbell-Switzer silver-pyridine method (**a**–**c**), Gallyas silver-iodide impregnations (**d**, **e**). Reproduced with permission from H Braak and E Braak, Temporal sequence of Alzheimer's disease-related pathology. Cerebral Cortex 1999;14:475–512

of resulting functional limitations in the brainstem nuclei with diffuse cortical projections. Provided assessment is made at different timepoints, such strategies could also facilitate prognostic estimates of the inter-individually different rates of

Table 9.1 Normal skill development and AD-related decline

Age range	Normal development and skills acquisition	Decline in AD and loss of skills
24–40 years	Performs complicated tasks	Can no longer perform complicated tasks
12–24 years	Manages finances	Can no longer manage finances
6–12 years	Selects clothing properly and dresses independently	Can no longer select clothing properly or dress independently
3–6 years	Can bath and toilet independently	Can no longer bath and toilet independently
2–3 years	Achieves fecal and urinary continence	Becomes incontinent for feces and urine
1–2 years	Can speak a few words and can walk	Speaks only a few words and can no longer walk
0.5–12 months	Can sit upright	Can no longer sit upright

the AD-related process. Pharmacologically, the present dominance of acetylcholinesterase inhibitors to treat the early impairment of the cholinergic magnocellular nuclei of the basal forebrain is understandable (Grantham and Geerts 2002; Smith et al. 2009; Medina and Ávila 2014b; but see also Tayab et al. 2012). Nevertheless, symptomatic relief and support of the diffusely projecting systems should be initiated as soon as possible and should be supplemented by substances that could also benefit the noradrenergic and serotonergic systems (Friedman et al. 1999; Dringenberg 2000; Grantham and Geerts 2002; Marien et al. 2004; Chalermpalanupap et al. 2013; Ramirez et al. 2014).

During stage III, the pathological process begins to move into the basal temporal neocortex (stage III: 276/2,366 cases = approximately 12 %) whereas, in stage IV it progresses further into limbic regions of the cortex (insula, subgenual and anterogenual regions, anterior cingulate gyrus) and temporal high order sensory association areas (stage IV: 75/2,366 cases = approximately 3 %). Cases at stages III–IV (351/2,366 cases = approximately 15 %) begin to occur in the third decade and increase in frequency after that up to the ninth decade (Figs. 9.10c, 9.13 stage III and IV, 9.16a). The majority of these cases show Aβ deposition (Table 9.3).

The ongoing deterioration of brainstem nuclei with diffuse cortical projections during stages III and IV leads to an increasing reduction of the noradrenergic, cholinergic, serotonergic, histaminergic, and dopaminergic input to the cerebral cortex. This influence normally modulates the activity level of cortical projection neurons in tandem with external and/or internal conditions. The cumulative damage inflicted on all of these sites exacts its toll on the complexity of cortical input and restricts the versatility with which cortical functions adapt to constantly changing demands. Such limitations ultimately pave the way for a reduction of higher cognitive functions (Chalermpalanupap et al. 2013; Reid and Evans 2013).

Against the backdrop of this generalized decline in cortical functioning, an additional more localized cortical pathology confined to the anteromedial temporal

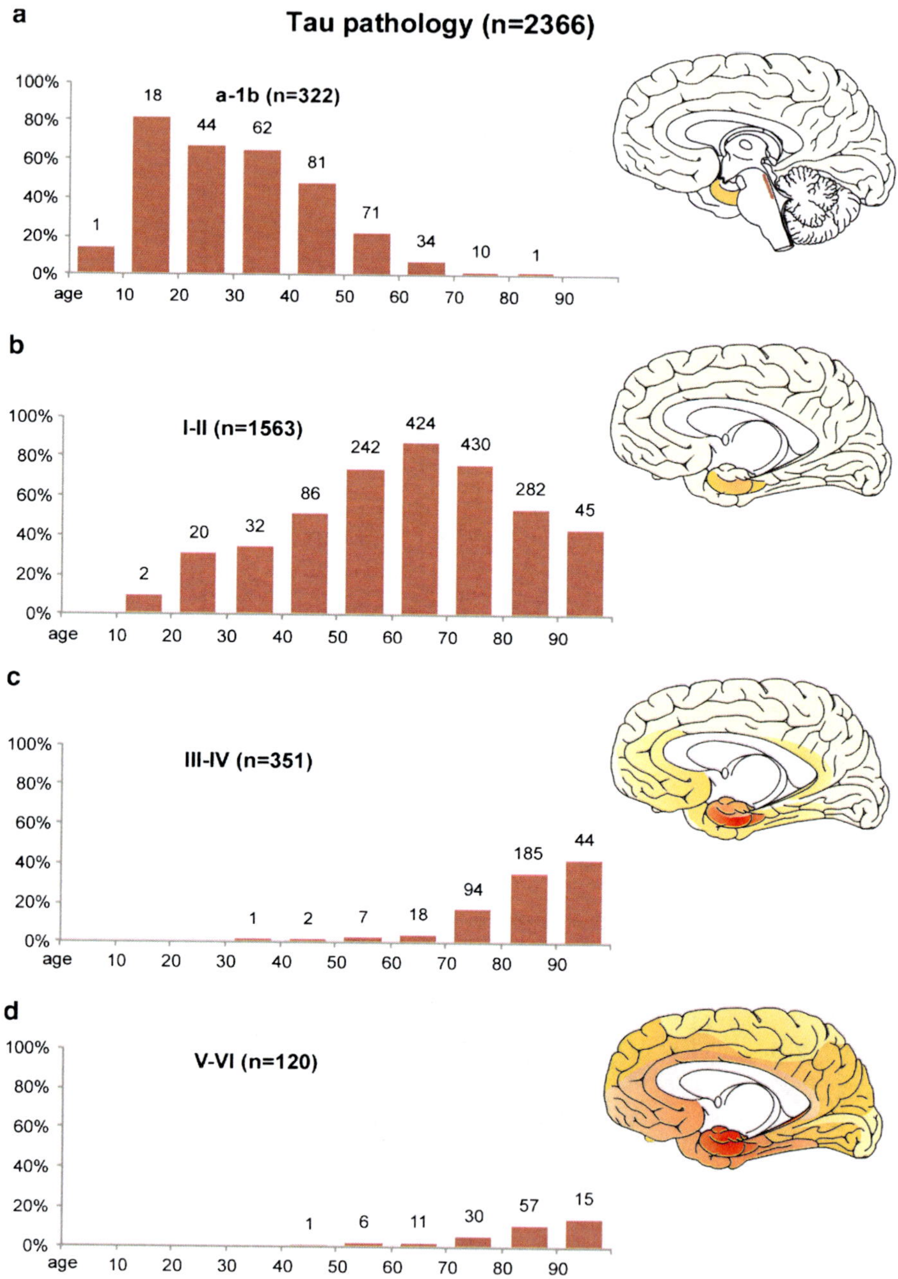

Fig. 9.10 Development of abnormal intraneuronal tau deposits in n = 2,366 non-selected autopsy cases according to decades (ages of the cohort 1–100). (**a**) The first graph illustrates the prevalence of AD-associated tau in non-argyrophilic lesions during stages a–1b (see also Fig. 7.4). (**b–d**) The following graphs show the prevalences of argyrophilic neurofibrillary lesions in NFT stages I–II (**b**), NFT stages III–IV (**c**), and NFT stages V–VI (**d**)

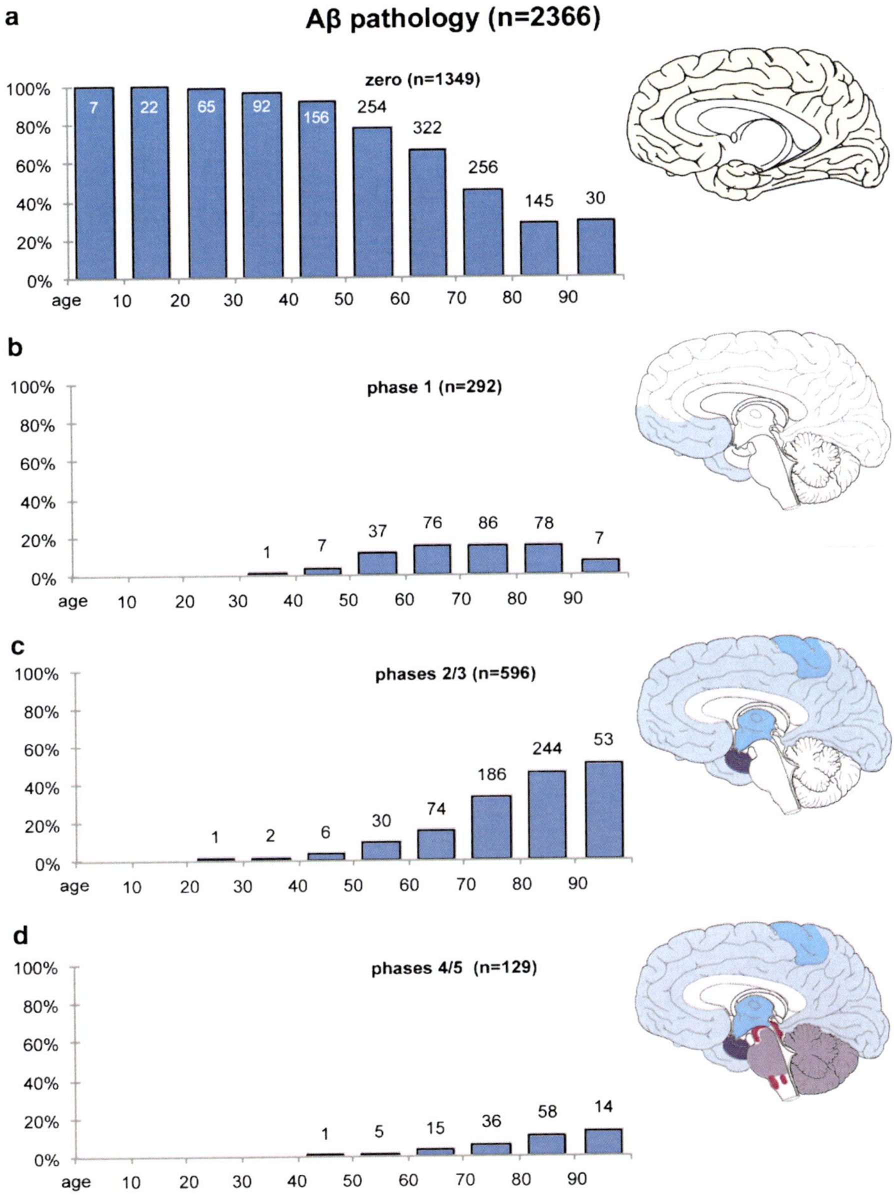

Fig. 9.11 Development of abnormal extracellular deposits of Aβ in n = 2,366 autopsy cases according to decades (ages of the cohort 1–100). (**a**) This graphs illustrates the relative prevalence of cases lacking Aβ deposits in the CNS. (**b–d**) These graphs show the prevalence of Aβ plaques corresponding to phase 1 deposition (**b**), phases 2–3 (**c**), and phases 4–5 (**d**)

lobe develops. This seriously impedes the data flow generated from high order processing fields of the neocortex and passing through the entorhinal region, hippocampal formation, ventral striatum, ventral pallidum, and mediodorsal

Table 9.2 Development of tau pathology according to decades (n = 2,366), including the ratio between females (n = 1,066) and males (n = 1,300)

Age	Zero	a–1b	I–II	III–IV	V–VI
0–9 n = 7	6 (2/4) 85.71 %	1 (0/1) 14.29 %	0 0 %	0 0 %	0 0 %
10–19 n = 22	2 (0/2) 9.09 %	18 (3/15) 81.82 %	2 (0/2) 9.09 %	0 0 %	0 0 %
20–29 n = 66	2 (2/0) 3.03 %	44 (18/26) 66.67 %	20 (10/10) 30.30 %	0 0 %	0 0 %
30–39 n = 95	0 0 %	62 (31/31) 65.26 %	32 (12/20) 33.68 %	1 (0/1) 1.05 %	0 0 %
40–49 n = 170	0 0 %	81 (31/50) 47.65 %	86 (38/48) 50.59 %	2 (0/2) 1.18 %	1 (1/0) 0.59 %
50–59 n = 326	0 0 %	71 (28/43) 21.78 %	242 (73/169) 74.23 %	7 (2/5) 2.15 %	6 (4/2) 1.84 %
60–69 n = 487	0 0 %	34 (9/25) 6.98 %	424 (149/275) 87.06 %	18 (7/11) 3.70 %	11 (6/5) 2.26 %
70–79 n = 564	0 0 %	10 (4/6) 1.77 %	430 (193/237) 76.24 %	94 (52/42) 16.67 %	30 (15/15) 5.32 %
80–89 n = 525	0 0 %	1 (0/1) 0.19 %	282 (149/133) 53.71 %	185 (118/67) 35.24 %	57 (34/23) 10.86 %
90–100 n = 104	0 0 %	0 0 %	45 (26/19) 43.27 %	44 (36/8) 42.31 %	15 (13/2) 14.42 %
Total n = 2,366	10 (4/6) 0.42 %	322 (124/198) 13.61 %	1,563 (650/913) 66.06 %	351 (215/136) 14.84 %	120 (73/47) 5.07 %

thalamus towards the prefrontal cortex (Fig. 9.16a). This input, which is vital for the prefrontal fields, can become curtailed to such an extent during stages III–IV that subtle signs of personality changes and impaired cognitive functioning come to light. Depending on additional factors, e.g., patient-specific neuronal reserve (Liberati et al. 2012) and preexisting comorbidities, the subcortical and cortical pathology that develops during stages III and IV can suffice to pave the way for some degree of clinically detectable cognitive impairment. Accordingly, the clinical histories of individuals at stage IV may refer to mild cognitive impairment (MCI), e.g., difficulties solving simple arithmetical or abstract problems, deficits of short-term memory (so-called 'working memory' deficits), and changes in personality ranging from apathy or depression to suspiciousness or irascible behavior and withdrawal from social contacts. MCI applies to individuals whose cognitive abilities are no longer normal but who are not demented (Petersen 2000, 2009, 2011; Petersen et al. 2001; Morris et al. 2001; Markesbery et al. 2006; Amieva et al. 2008; Sabbagh et al. 2010; Mayeux 2010; Albert et al. 2011; Ewers et al. 2012; Mufson et al. 2012; Rentz et al. 2013).

Moreover, stage IV lesions also systematically erode the superordinate autonomic relay centers that regulate respiration, heart rate, blood pressure, respiration, and gastrointestinal motility, among other functions, and these centers include the insular, subgenual and anterogenual regions, anterior cingulate areas as well as the

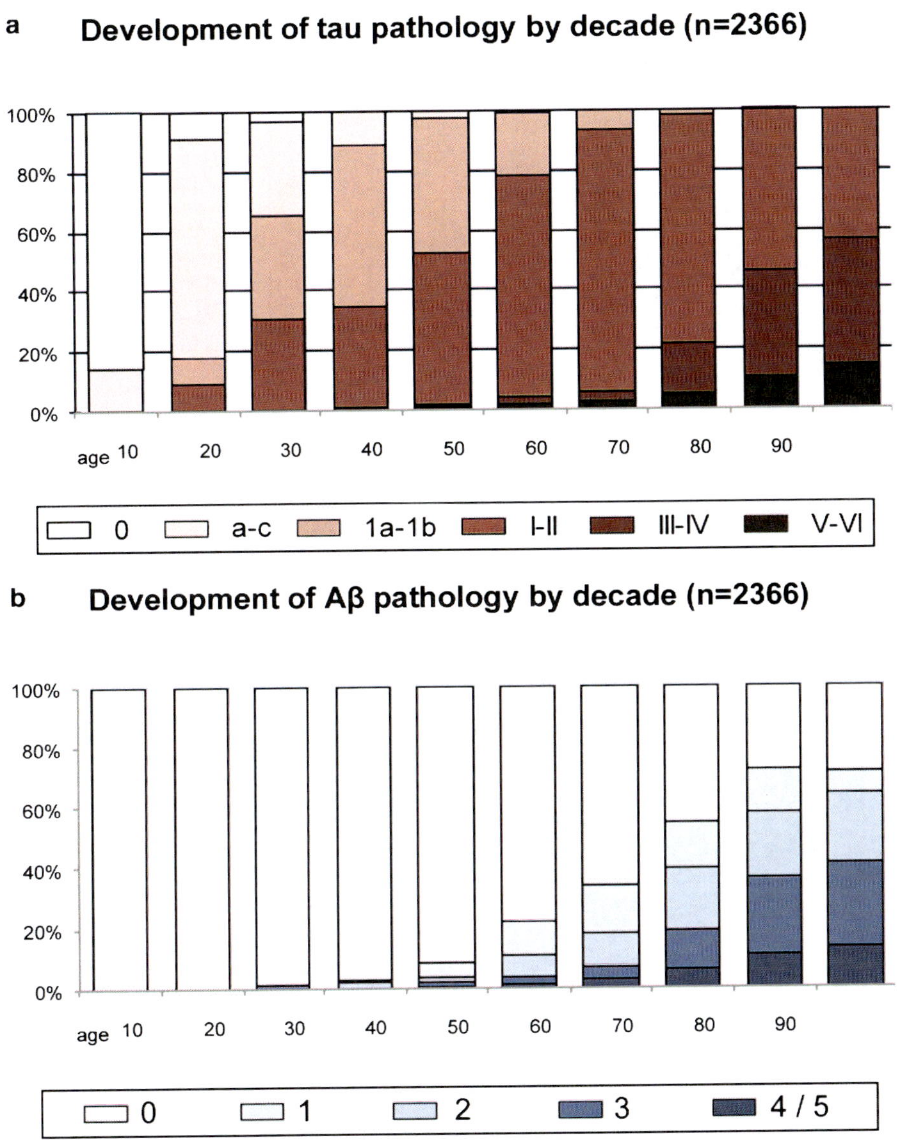

Fig. 9.12 Summary diagrams of the prevalence of tau stages alongside of phases of Aβ deposition according to decades (n = 2,366). *White columns* indicate an absence of tau pathology (**a**) and Aβ deposits (**b**). (**a**) Columns in *red* shading indicate the relative frequency of cases with all stages of intraneuronal lesions. (**b**) Columns in *blue* shading indicate the relative frequency of cases showing various phases of plaque-like Aβ deposition. Note the relatively late appearance of Aβ plaques

central subnucleus of the amygdala (Chu et al. 1997; Royall 2008; Royall et al. 2008). The structural and functional integrity of involved autonomic centers are necessary for maintaining the sympathetically mediated increase in heart rate

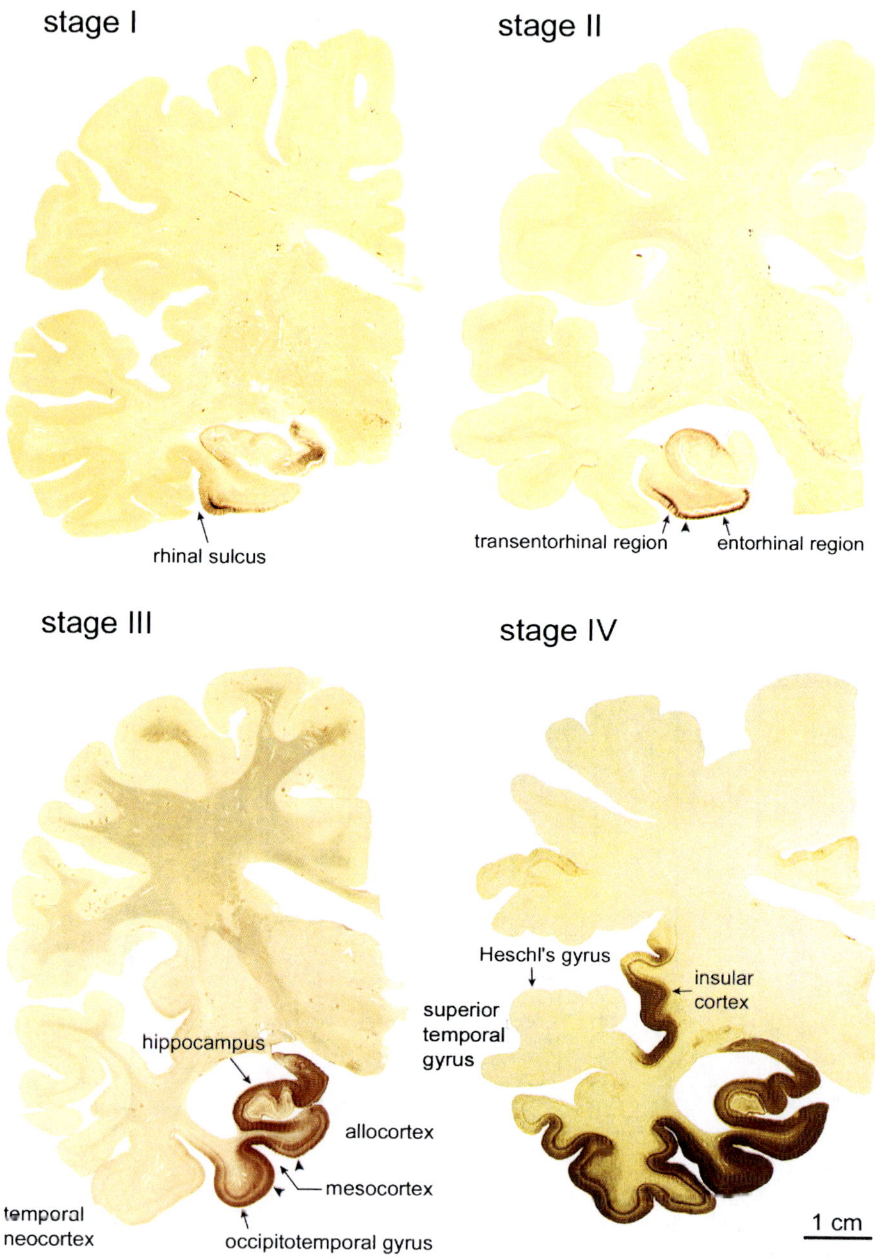

Fig. 9.13 NFT stages I–IV in AT8-immunostained 100 μm hemisphere sections. *Stage I*—Involvement is slight and all but confined to the transentorhinal region (cognitively intact 80-year-old female). *Stage II*—Additional immunoreactivity occurs in layer pre-α of the entorhinal region. The layer gradually sinks into a deeper position in the transentorhinal region. Thus, the border between both regions is clearly recognizable (*arrowhead*). The lesions also advance

during physical exertion or emotional stress, the resting level for the gain of the cardiac component of the baroreflex response, and appropriate sudomotor responses of the skin to emotional stimuli (Zakrzewska-Pniewska et al. 2012; Meel-van den Abeelen et al. 2013; Femminella et al. 2014).

The autonomic nervous system outside the CNS (i.e., PNS/ENS) together with the sympathetic and parasympathetic centers in the medulla oblongata and spinal cord do not consistently become involved in the AD process. Thus, basic autonomic functions remain intact, whereas the impairment within the central autonomic network is confined to higher levels of modulating regions (amygdala, limbic nuclei of the thalamus, limbic areas of the cerebral cortex) (Benarroch 1993) (Fig. 6.10). It would represent a significant advance, if one could learn to utilize dysfunctional phenomena of the superordinate autonomic relay centers for diagnostic purposes because the networks become involved early (stage III: central subnucleus of the amygdala) and, by stage IV, symptoms should be detectable. The necessary diagnostic tools still have to be developed but they would complement those already available for diagnosing MCI (Mufson et al. 2012), thereby making it possible to strengthen the clinical impression.

The prevalence of the late stages V–VI rises with age (stage V: 100/2,366 cases = approximately 4 %, stage VI: 20/2,366 cases = about 1 %; altogether 120/2,366 cases = approximately 5 %) (Figs. 9.10d, 9.14, and 9.16b). Individuals with a growing burden of pathology generally display a shift toward higher age categories (Fig. 9.10d). The resulting extensive over-regional impairment of the neocortex usually leads beyond the threshold to manifest symptoms, thereby making the clinical diagnosis of AD possible (Nelson et al. 2007, 2009; Dolan et al. 2010; Sabbagh et al. 2010). In general, the degree of the intraneuronal tau pathology correlates with the presence and severity of the clinical symptoms of sporadic AD (Grober et al. 1999; Gold et al. 2000; Giannakopoulos et al. 2003; Tiraboschi et al. 2004; Nelson et al. 2007: Delacourte 2008; Dolan et al. 2010). Furthermore, all cases at stages V and VI have Aβ deposition corresponding to phase 2 or higher (Table 9.3) (Braak and Braak 1997b; Thal et al. 2002). The Aβ

Fig. 9.13 (continued) slightly into the hippocampal formation (cognitively intact 80-year-old male). *Stage III*—The lesions in the hippocampal formation worsen. Entorhinal layer pre-α and, additionally, layer pri-α of the deep layers become heavily involved. The tau pathology progresses through the transentorhinal region into the adjoining high order association areas of the basal temporal neocortex. *Arrowheads* mark the extent of the transentorhinal region (mesocortex). The lesions usually do not go beyond the occipito-temporal gyrus laterally and the lingual gyrus posteriorly (90-year-old female). *Stage IV*—A large portion of the insular cortex becomes involved. The lesions in the high order sensory association cortex of the temporal lobe now extend into the medial temporal gyrus without entering the superior temporal gyrus. The primary fields of the neocortex (see transverse gyrus of Heschl) and, to a large extent, also the premotor and first order sensory association areas of the neocortex remain intact (82-year-old female AD patient). Reproduced with permission from H Braak et al., Vulnerability of cortical neurons to Alzheimer's and Parkinson's diseases. J Alzheimers Dis 2006;9(3 Suppl):35–44

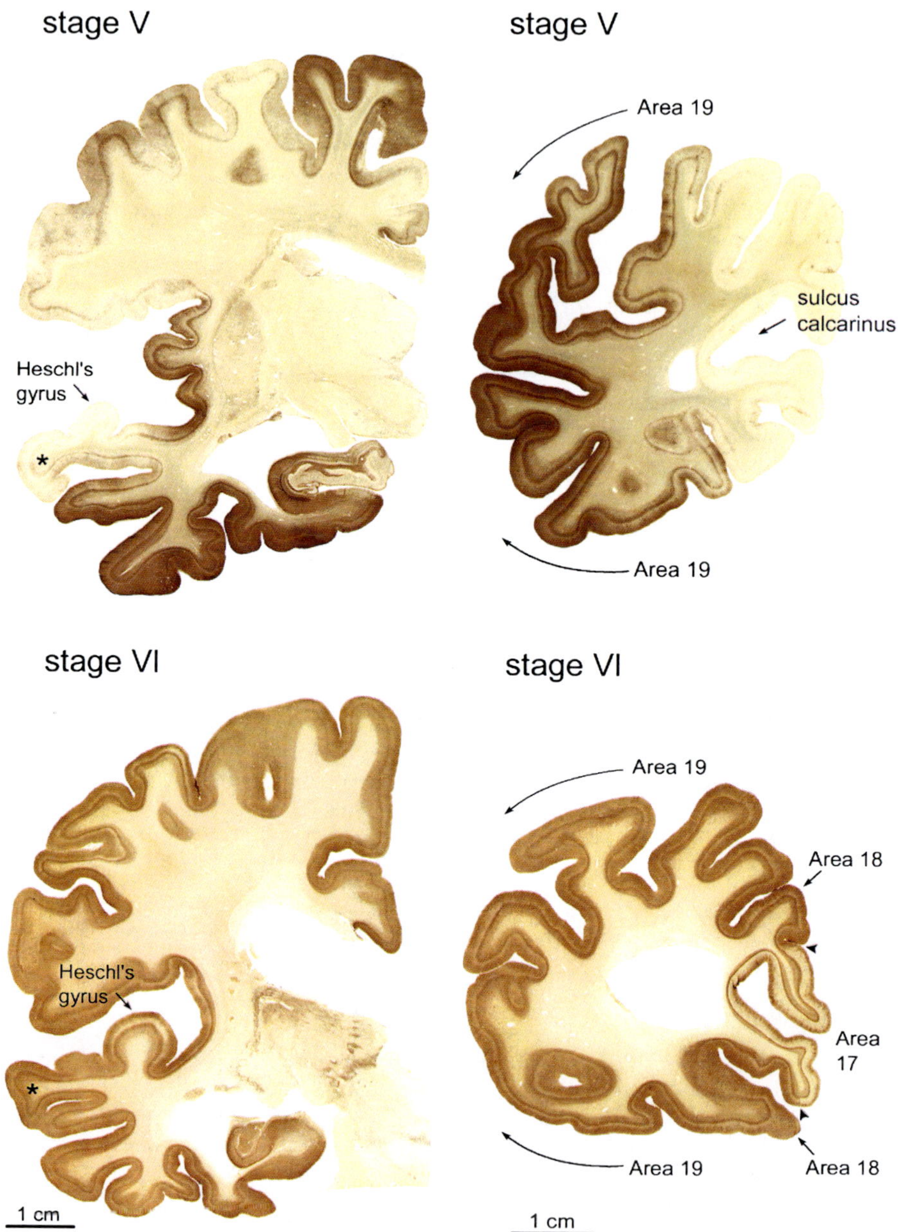

Fig. 9.14 NFT stages V and VI in AT8-immunostained 100 μm sections. *Stage V*—Tau pathology appears in the superior temporal gyrus and encroaches, to a mild degree, upon premotor areas and first order sensory association fields. In the occipital lobe (at *right*) the peristriate region shows varying degrees of involvement, and lesions occasionally can be seen even in the parastriate area (90-year-old female AD patient). *Stage VI*—Severe pathology is detectable even in the parastriate and striate areas (primary visual cortex). The conspicuous differences between occipital tissue

plaque load, however, does not correlate with the severity of clinical AD symptoms (Giannakopoulos et al. 2003; Nelson et al. 2007, 2009).

The intellectual faculties do not all deteriorate to the same extent. In the CNS, it is striking how long all of the subunits within the well-myelinated auditory system resist the progress of the pathological process. Despite this, AD patients eventually present with deficits in sound-source localization and auditory frequency discrimination (Kurylo et al. 1993; Sinha et al. 1993; Gates et al. 2002; Dugger et al. 2011), but because cortical components of the auditory system remain nearly intact, patients can be reached for a long time via auditory stimuli and, therefore, can be motivated to dance and perform rhythmic gymnastics.

Clinical histories of individuals with stages V or VI pathology generally note the presence of fully-developed AD with severe dementia, i.e., marked memory loss as well as loss of executive functions (Fig. 9.16b). Severe autonomic dysfunction reflects the extensive destruction of limbic circuit centers, and these persons become incontinent for urine and feces, they no longer can dress themselves independently, and they fail to recognize persons once familiar to them (spouse, children). Later, they can no longer walk, sit up, or hold up their head unassisted. Increasing stiffness of the large joints owing to lack of exercise or use produces contractures of the extremities and, ultimately, immobility. Primitive reflexes that normally are present only in infants reappear in the latest AD phase (Table 9.1) (Franssen et al. 1993; Reisberg et al. 1999; Nelson et al. 2007; Dolan et al. 2010; Sabbagh et al. 2010). The time period that elapses between the onset of clinical symptoms and death averages approximately eight to nine years (Zanelli et al. 2009).

Cases with tau pathology but without Aβ plaque deposition, or cases with tau pathology displaying only a very few plaques (phases 0 or 1) are shown in Tables 7.2 and 9.3. The number of these cases decreases considerably between NFT stages III and IV. The earliest Aβ deposits begin to develop in stages 1a–1b (Table 7.2). Advanced AD cases (NFT stages V or VI) always are accompanied by phase 2 of Aβ plaque deposition or higher (Table 9.3). Tables 7.2 and 9.3 do not give occasion to postulate the existence of a non-AD variant of aging, i.e., a tauopathy devoid of Aβ deposits, alongside of the tau pathology characterizing the AD-related pathological process (see Sect. 9.3). At the same time, this clinically bland but remarkably large group of tau-only individuals (1,258 of 2,366 cases) bears observation and closer study to clarify, among others, why Aβ deposits develop relatively late during the disease course.

The average age (arithmetic mean) of cases at stages a-VI increases with age between stages I and IV when argyrophilic (Gallyas-positive) lesions are taken into account, and increases, when AT8-ir pathology is assessed, from no tau lesions up

---

Fig. 9.14 (continued) sections at NFT stage V and stage VI can and should be used for the diagnostic differentiation between the final two stages (70-year-old female AD patient). Reproduced with permission from H Braak et al., Vulnerability of cortical neurons to Alzheimer's and Parkinson's diseases. J Alzheimers Dis 2006;9(3 Suppl):35–44

Table 9.3 Comparison of stages III–VI of AD-associated tau lesions lacking Aβ deposits (n = 81) and phases 1–5 of Aβ deposition (n = 390)

Aβ phases	Tau stage III	Tau stage IV	Tau stage V	Tau stage VI	Tau stages III–VI
Zero	76	5	0	0	81
1	35	2	0	0	37
2	54	11	3	0	68
3	100	39	24	1	164
4	10	18	36	15	79
5	1	0	37	4	42
Total	276	75	100	20	471

Table 9.4 Mean age of cases with no pathology and with lesions corresponding to tau stages a–VI evaluated using both Gallyas silver staining and AT8 immunoreactions

Tau stage	n (Gallyas)	Mean age	n (AT8)	Mean age
Zero	332	42.4	10	9.7
a–c	–	–	57	27.4
1a–1b	–	–	265	46.9
I	980	63.6	612	59.9
II	583	75.3	513	70.3
III	276	80.1	504	76.5
IV	75	83.1	191	80.6
V	100	80.7	132	81.8
VI	20	74.7	82	80.0

to and including stage V. The average age falls off somewhat at higher stages (Table 9.4). This trend could be an indication that the pathological process accelerates in its final phase. The timespan between cases with early stages a–c (AT8) and those with late stages V/VI (AT8) amounts to approximately fifty years (Table 9.4, compare with data in Table 9.7).

Table 9.5 shows the number of cases with evaluation of the same stages in both Gallyas-silver stained (G) material and AT8 (A) immunoreactions (G = A: 97 of 276 stage III cases = approximately 35 %; 9 of 75 stage IV cases = approximately 12 %; 42 of 100 stage V cases = approximately 42 %; and 20 of 20 stage VI cases = 100 %). However, cases do exist that show a difference of one stage (153 of 276 stage III cases = approximately 55 %, 64 of 75 stage IV cases = approximately 85 %; 58 of 100 stage V cases = 58 %) and, sometimes, even a difference of two stages (26 of 276 stage III cases = approximately 9 % and 2 of 75 stage IV cases = about 3 %).

The course of the AD process depicted in the diagrams appears for the most part – at least in the initial phase – to progress at the same rate. Cases with greater variations between the AT8 and Gallyas staging diagnoses are more prevalent in older age groups (Table 9.5). Once again, it is possible that the pace of the pathological process during the late stages gains momentum (Tables 7.3 and 9.5; see also Table 9.4).

Table 9.5 Differences between Gallyas and AT8 scores (NFT stages III–VI cases)

Age	n=G III	G=A	=III+1	=III+2	1+2	=%	n=G IV	G=A	=IV+1	=IV+2	1+2	=%	n=G V	G=A	=V+1	=%	n=G VI=A VI
30–39	1	1	0	0	0	0	0	0	0	0	0	0	0	0	0	0	0
40–49	2	2	0	0	0	0	0	0	0	0	0	0	0	0	0	0	1
50–59	6	3	3	0	3	50	1	1	0	0	0	0	4	3	1	25	2
60–69	16	9	6	1	7	44	2	1	1	0	1	50	9	4	5	56	2
70–79	80	40	34	6	40	50	14	2	12	0	12	86	23	9	14	61	7
80–89	140	38	85	17	102	72	45	3	40	2	42	93	49	21	28	57	8
90–100	31	4	25	2	27	87	13	2	11	0	11	85	15	5	10	67	0
Total	276	97	153	26	179	64	75	9	64	2	66	88	100	42	58	58	20

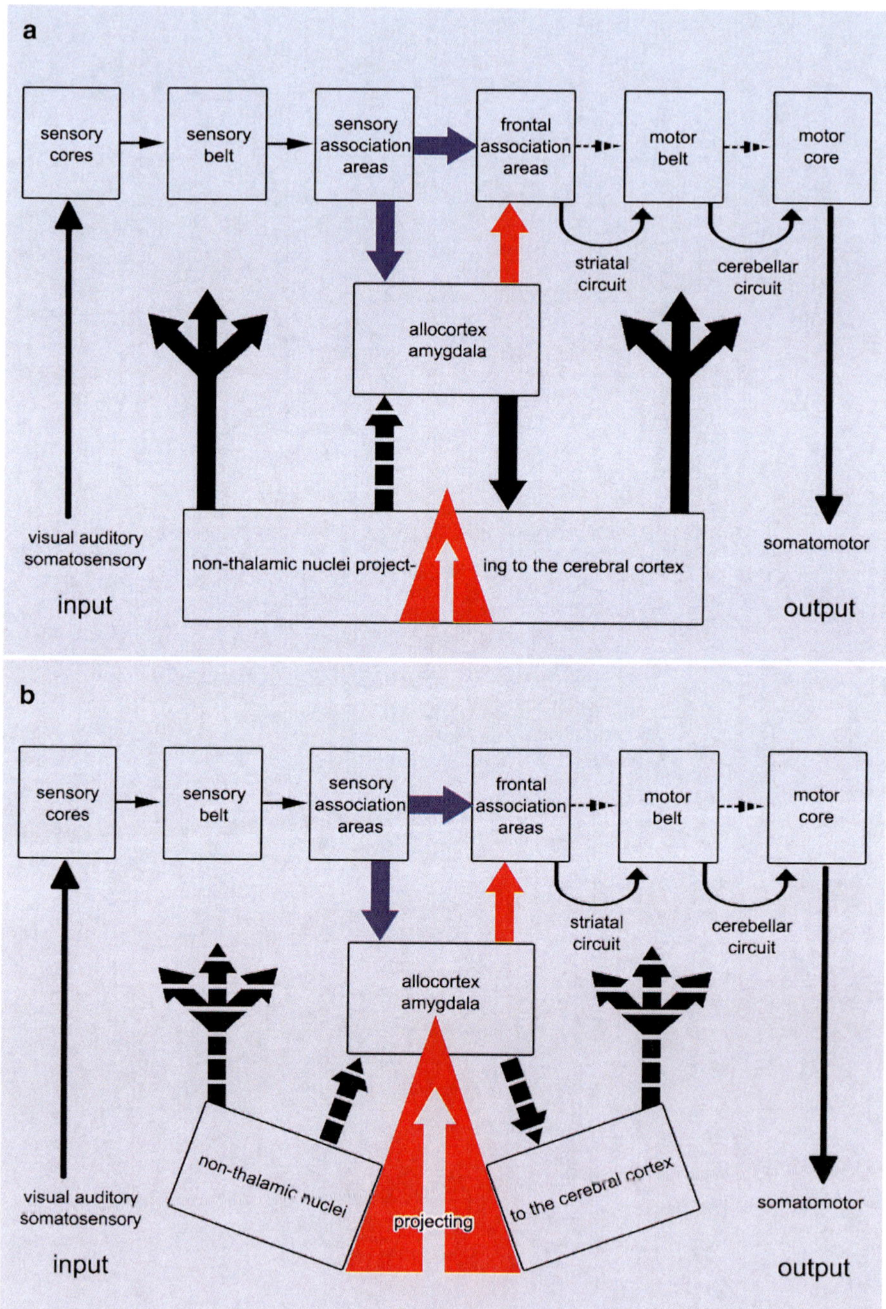

Fig. 9.15 AD is a 'disconnection syndrome' part 1 (Kemper 1984; Hyman et al. 1990; Delbeuck et al. 2003; Reid and Evans 2013). (**a**) AD-related tau pathology commences in the noradrenergic locus coeruleus and then goes on to involve the other brainstem nuclei (stages a–c: *red* wedge) that send diffuse ascending projections to the cerebral cortex (*thick black arrows* at *left/right*). These brainstem nuclei influence cortical processing and modulate the activity level of cortical projection

Table 9.6 Development of Aβ plaques according to decades (n = 2,366), including the ratio between females (n = 1,066) and males (n = 1,300)

Age	Zero	Phase 1	Phases 2 + 3	Phases 4 + 5
0–9 n = 7	7 (2/5) 100 %	0 0 %	0 0 %	0 0 %
10–19 n = 22	22 (3/19) 100 %	0 0 %	0 0 %	0 0 %
20–29 n = 66	65 (30/35) 98.48 %	0 0 %	1 (0/1) 1.56 %	0 0 %
30–39 n = 95	92 (41/51) 96.84 %	1 (1/0) 1.05 %	2 (1/1) 2.11 %	0 0 %
40–49 n = 170	156 (64/92) 91.76 %	7 (2/5) 4.12 %	6 (3/3) 3.53 %	1 (1/0) 0.59 %
50–59 n = 326	254 (84/170) 77.91 %	37 (12/25) 11.35 %	30 (8/22) 9.20 %	5 (3/2) 1.53 %
60–69 n = 487	322 (110/212) 66.12 %	76 (21/55) 15.61 %	74 (32/42) 15.20 %	15 (8/7) 3.08 %
70–79 n = 564	256 (101/155) 45.39 %	86 (35/51) 15.25 %	186 (108/78) 32.98 %	36 (20/16) 6.38 %
80–89 n = 525	145 (73/72) 27.62 %	78 (41/37) 14.86 %	244 (153/91) 46.48 %	58 (34/24) 11.05 %
90–100 n = 104	30 (20/10) 28.85 %	7 (4/3) 6.73 %	53 (38/15) 50.96 %	14 (13/1) 13.46 %
Total n = 2,366	1,349 (528/821) 57.02 %	292 (116/176) 12.34 %	596 (343/253) 25.19 %	129 (79/50) 5.45 %

The columns in Fig. 9.11a–d show the prevalences of all phases of Aβ deposition for all 2,366 cases. The same data (Aβ phases) separated according to gender can be found in Table 9.6. Primitive Aβ plaques generally begin to develop in the fourth decade and reach a maximum in the eighth decade. Beyond that point, phase 1 cases decline again in frequency and are replaced by cases with more advanced Aβ deposition (Fig. 9.11b). Aβ plaques then appear in previously uninvolved regions, reaching phases 2–3 (Fig. 9.11c) and, in some instances, phases 4 and 5 (Fig. 9.11d). Once again, the diagrams show the proposed continuum (Fig. 9.11), with a shift towards higher age categories, beginning with the occurrence of Aβ plaques in basal portions of the temporal neocortex and peaking in the late phases. Early phases of Aβ deposition are seen predominantly in younger individuals, whereas the more advanced phases occur at higher ages, i.e., the mean of Aβ deposition phases increases with age. As such, age is a risk factor for

Fig. 9.15 (continued) neurons in response to external and/or internal conditions and input/output (*slender black arrows* at far *left/right*). (**b**) Subsequently, the pathology develops essentially in the transentorhinal and entorhinalen regions (stages 1a,1b and NFT stages I–II: *red* wedge). The lesions target the host neurons of the perforant pathway and thereby reduce the transfer of neocortical date heading towards the hippocampal formation

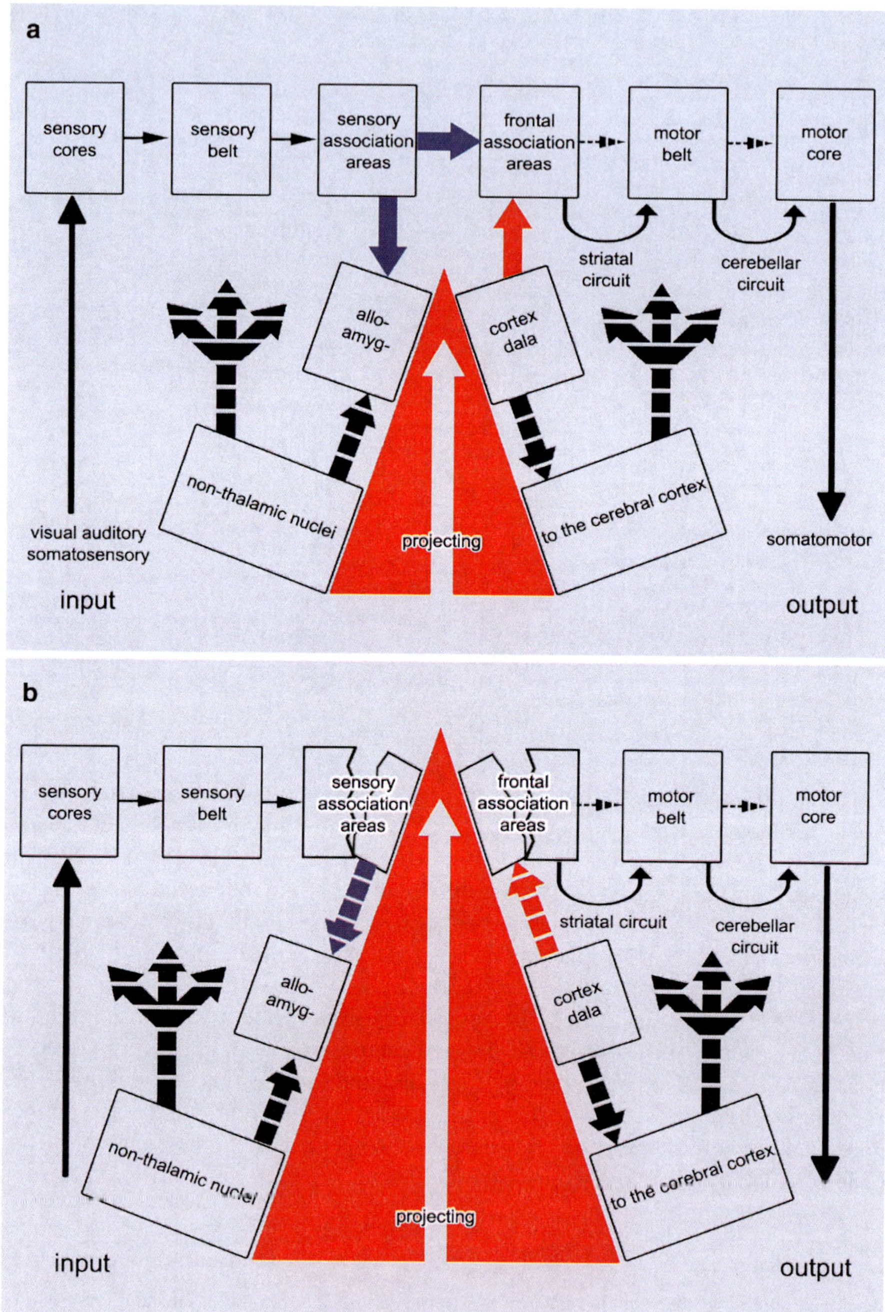

Fig. 9.16 AD is a 'disconnection syndrome' part 2. (**a**) Progressive deterioration of nerve cells results in the nearly complete destruction not only of the projections from the neocortex to the hippocampal formation but also the projections from the hippocampal formation to the neocortex—both the afferent and efferent arms of the limbic circuit (NFT stages III and IV: *red* wedge). (**b**) The pathological process then expands throughout the neocortex and, in this manner,

Aβ deposition. As regards average age data, the span between cases at phase 1 and those at phase 5 amounts to approximately 8 years (Table 9.7).

The summary Fig. 9.12 illustrates the proposed continuum of the intraneuronal pathology ranging from non-fibrillar material in axons (subcortical stage a) to the extensive regional distribution pattern that is characteristic of advanced AD (NFT stage VI) (Braak et al. 2011). The extraordinarily long time span over which the lesions develop, combined with the observation that they develop in a very large proportion of the human population does not detract from their insidious nature or imply that the lesions are 'normal' (see Chap. 2.1). If stages a–c, 1a,1b, and I–II are incorporated into the natural history of the tau pathology, it can be postulated that the preclinical phase of AD is nearly a lifelong process. Yet, this also implies that the AD process offers a much larger window of opportunity for disease-modifying interventions than previously imagined (Braak and Del Tredici 2012).

Figure 9.12a, b juxtapose the development of tau inclusions and Aβ plaques for all of the cases shown previously. Only a proportion of all cases exhibit Aβ plaque deposition and, notably, all of the Aβ-positive cases have tau pathology. Tau inclusions corresponding to stages a–c and 1a–b develop approximately two decades prior to the appearance of Aβ plaques (Braak et al. 2011).

Summaries of all the cases permit the compilation of five tau groups (a–c, 1a–1b, I–II, III–IV, V–VI) and three Aβ groups (phase 1, phases 2–3, phases 4–5). Of the 15 potential combinations among all 8 groups, only 10 actually emerge (Table 9.8). No case in the entire cohort displays Aβ deposits in the absence of tau pathology.

Finally, Table 9.9 provides data regarding the not uncommon coincidence of AD- and PD-associated pathologies. The cohort includes n = 187/2,366 cases with PD-related synucleinopathy. These cases, in turn, can be subdivided into three groups (PD stages 1–2, PD stages 3–4, PD stages 5–6) (Braak and Del Tredici 2009) and, again, five tau groups (a–c, 1a–1b, I–II, III–IV, V–VI). This time, of the fifteen possible combinations, eleven emerge (Table 9.9). It should be emphasized that no case displays PD-associated pathology in the absence of AD-related lesions. All cases with PD-related synucleinopathy also have AD-associated pathology. Judging from the results shown in Table 9.9, the presence of at least stages 1a or 1b (i.e., cortical involvement with non-argyrophilic tau lesions) is needed for the development of PD-associated pathology. The consequences or implications of the tau lesions for the progression of the PD process are still unknown (Halliday

Fig. 9.16 (continued) ultimately destroys the limbic circuit: both the afferences originating from high order sensory association areas and also the vital efferences that decisively influence the prefrontal neocortex via the vental striatum, ventral pallidum, and the dorsomedial nucleus of the thalamus. The cumulative effect of the tau pathology attenuates the general input to the cerebral cortex and, by severing the limbic system from the cortex (stages V and VI: *red* wedge and *dashed arrows*), ultimately prevents the cortex from adapting to constantly changing demands involved in executive functioning. In the final analysis, this 'disconnection' of cortical projection neurons from the superordinate centers of the limbic system paves the way for the development of dementia in AD

Table 9.7 Mean age of cases with no pathology and with Aβ deposition corresponding to phases 1–5

Phase	n (Aβ)	Mean age
Zero	1,344	60.0
1	293	71.3
2	340	75.7
3	260	79.7
4	87	79.0
5	42	79.1

Table 9.8 Number of cases in various combinations of tau stages with phases of Aβ deposition

AD-associated tau stages	Aβ phase 1	Aβ phases 2/3	Aβ phases 4/5	Total
a–c	0	0	0	0
1a–1b	12	9	0	21
I–II	244	359	8	611
III–IV	37	204	29	270
V–VI	0	28	92	120
Total	293	600	129	1,022

Table 9.9 Number of cases showing various combinations of tau stages with PD-associated α-synuclein pathology

AD-associated tau stages	PD stages 1/2	PD stages 3/4	PD stages 5/6	Total
a–c	0	0	0	0
1a–1b	1	5	0	6
I–II	44	54	23	121
III–IV	9	14	10	33
V–VI	8	7	12	27
Total	62	80	45	187

et al. 2014). The PD-associated lesions generally begin to develop in the fourth decade and then increase in higher age groups.

Chapter 10
Final Considerations

The present exposition of the AD-related pathological process rests upon some basic assumptions:

(1) The AD process is a uniquely human condition involving specific nerve cell types of the CNS and it progresses along a biological continuum.
(2) Tau pathology with or without Aβ deposition in cognitively intact individuals are not 'normal age changes' but represents a premorbid state.
(3) AD-associated pathology causes a functional decline of vulnerable nerve cells rather than global neuronal loss.
(4) Physical (i.e., synaptic) contacts between susceptible projection cells of involved regions play a key role in the pathogenesis of sporadic AD.
(5) The AD process is not Aβ-dependent.

Because AD is a disorder of the *human* CNS for which no truly suitable animal model exists (Reid and Evans 2013), the ongoing trend away from autopsy-controlled diagnosis is irresponsible, unwarranted, and even detrimental to the AD field. Many of the unsolved questions and issues related to the neuropathology of AD can be addressed effectively in tissue from autopsy-controlled prospective studies performed on cohorts that include individuals from normal aging populations (Morrison and Hof 1997; Nelson et al. 2011). Among those requiring answers, several mentioned in this monograph are summarized here as follows:

(1) Why does the AD-related pathological process develop almost exclusively within vulnerable nerve cell types of the CNS, although neurons in the PNS and ENS have normal tau and the amyloid precursor protein (APP)?
(2) Do molecular and conformational changes make oligomeric tau toxic to crucial cellular functions?
(3) Does the induction of the pathological protein misfolding process entail mechanisms and pathways that are different from those regulating its propagation?

H. Braak, K. Del Tredici, *Neuroanatomy and Pathology of Sporadic Alzheimer's Disease*, Advances in Anatomy, Embryology and Cell Biology 215,
DOI 10.1007/978-3-319-12679-1_10

(4) Can conformational variants influence the degree of tau pathogenicity as well as the rate of the pathological process?
(5) Why does non-argyrophilic abnormal tau material develop in both the axon and somatodendritic compartment of vulnerable nerve cell types?
(6) Why does argyrophilic abnormal tau material only develop in the cell soma (NFTs) and dendrites (NTs) but not in the axon of involved neurons?
(7) Which neuronal types in the human CNS produce and secrete Aβ?
(8) At which specific cellular sites is Aβ released into the ISF?
(9) How do Aβ deposits develop from expansive (i.e., cloud-like) and unsharply defined structures into compact primitive or cored plaques?
(10) Why do Aβ plaques not develop in the pallidum, lateral tuberal nuclei, and lateral mamillary nucleus?
(11) Why does the spinal cord develop tau pathology and CAA without Aβ plaques?
(12) Can differences in the individual 'reserve capacity' influence the rate of the pathological process?

The AD-related pathological process can begin in childhood or in young adulthood but it exists in nearly every individual and continues into old age. The topographic distribution of the pathology reveals that the process follows a stereotypic pattern. The rate of progression, however, shows considerable inter-individual differences, and this explains why only a relatively small number of persons become demented. Most individuals do not develop a sufficient degree of AD-associated pathology during the course of a lifetime to manifest a clinically recognizable symptomatology of MCI or AD.

In addition to neuroimaging and biomarker improvement, more refined diagnostic instruments developed to acquire data pertaining to dysfunctions of the diffusely projecting nuclei would help to identify the presence of the pathological process closer to the time of its inception. At present, the clinical diagnosis of AD is usually made only when the pathological process has reached NFT stage V or VI, which is much too late.

Initiation of therapeutic interventions during the end-stages of the AD pathological process is futile because available anti-dementia drugs do not arrest, reverse, or even modify the tau and Aβ pathologies (Molnar et al. 2009; Maarouf et al. 2010). Most efforts at developing effective strategies aimed at preventing abnormal protein aggregation, reducing Aβ/tau levels, or removing the protein accumulations have been unsuccessful (Bulic et al. 2010; Gong et al. 2010; Iqbal and Grundke-Iqbal 2011; Wagner et al. 2013; Lou et al. 2014). Mechanisms still to be elucidated are, for example, not only those of variant tau conformers (Sanders et al. 2014) but also those influencing the degree of myelination of vulnerable projection neurons (Ullén 2009; Fields 2011). Nerve cell activity provides the physiological stimulus for oligodendroglia cells to produce and sustain the myelin sheath (Ullén 2009). The more active the nerve cell, the thicker the myelin sheath becomes during the differentiation process and, presumably, the better protected the neuron is against the AD process. As such, it makes sense to seek means of slowing the rate at which the pathological process progresses, including augmentation of the individually variable 'cognitive reserve' (Savica and Petersen 2011; Erickson et al. 2012; Meng and D'Arcy 2012; Barulli and Stern 2013; Serrano-Pozo et al. 2013; Xu et al. 2014) because even a slight delay in the rate of progression would mean a sharp decrease in the number of patients crossing the clinical threshold to MCI and AD. The aim would be to alter the postnatally ongoing myelination process to such an extent that quantifiable increases in the degree of myelination would correspond to greater resistance on the part of vulnerable nerve cells (Stern 2002, 2009; Bengtsson et al. 2005; Schlaug et al. 2003; Draganski and May 2008; Craik et al. 2010; Gärtner et al. 2013; Freedman et al. 2014).

Chapter 11
Technical Addendum

Brains obtained at autopsy—following precautions for handling tissue of prion diseases—are fixed by immersion in 10 % formalin (4 % aqueous solution of HCHO) for 1 week or longer. Remove the brainstem and cerebellum and cut the brain mid-sagittally. Take one hemisphere and remove the meninges to uncover the rhinal sulcus and the calcarine fissure. The staging procedure for AD requires evaluation of a few sections from a minimum of (A) two blocks of cortical tissue and (B) one brainstem block. The first block includes anteromedial portions of the temporal lobe (frontal section at the level of the mamillary bodies), encompassing uncal portions of the hippocampus, the parahippocampal gyrus, and adjoining temporal gyri (Braak et al. 2006a). The cutting line typically runs through the rhinal sulcus, including the transentorhinal and entorhinal regions (Fig. 11.1a, cutting line 1). The tissue on one side of the cut can be used for conventional paraffin embedding (70-100 m sections; see Feldengut et al. 2013), and that on the other side for frozen sections or polyethylene glycol (PEG) embedding (100-400 μm sections). This block of temporal lobe tissue is essential for assessment of stages 1a, 1b, and NFT stages I–IV, and it contains the transentorhinal and entorhinal regions. A section thickness of 70-100 μm meets the demands of low power light (stereo)microscopy and greatly facilitates recognition of the lesional distribution pattern (Fig. 11.1).

The conventional block of hippocampal tissue usually is removed at the level of the lateral geniculate nucleus (Fig. 11.1a, cutting line 2). However, at this latitude the adjoining parahippocampal gyrus often includes only posterior remnants of the transentorhinal and entorhinal regions. Since evaluation of these regions is essential to the staging procedure, care should be taken to remove the temporal block a few millimeters anterior to the level of the lateral geniculate nucleus (Braak et al. 2006a).

The second block needed for staging is removed preferably halfway between the occipital pole and the junction of the parieto-occipital sulcus with the calcarine fissure (Fig. 11.1a, cutting line 3). The cut is oriented perpendicular to the calcarine

H. Braak, K. Del Tredici, *Neuroanatomy and Pathology of Sporadic Alzheimer's Disease*, Advances in Anatomy, Embryology and Cell Biology 215,
DOI 10.1007/978-3-319-12679-1_11

fissure. The size of the block can be reduced along the dashed line to fit conventional cassettes but should include the cortex covering the lower bank of the calcarine fissure and the adjoining basal occipital gyri. It shows occipital association areas, the parastriate area, and an easily delineable primary field, the striate area. This occipital block is essential for recognition of neurofibrillary stages V–VI.

The brainstem block needed for the diagnosis of the tau process is a section through the locus coeruleus and dorsal raphe nucleus at about the level 1 in Fig. 11.1b. The other levels are required for additional diagnoses of neurodegenerative diseases other than AD.

Specific antibodies for the pathologic proteins are available. However, their application often requires special fixation methods and/or antigen retrieval methods which influence the result of the immunoreaction to a variable extent. Dye staining techniques (congo red, thioflavine S) are rather insensitive, while conventional silver methods show both normal and pathological components (Bielschowsky,

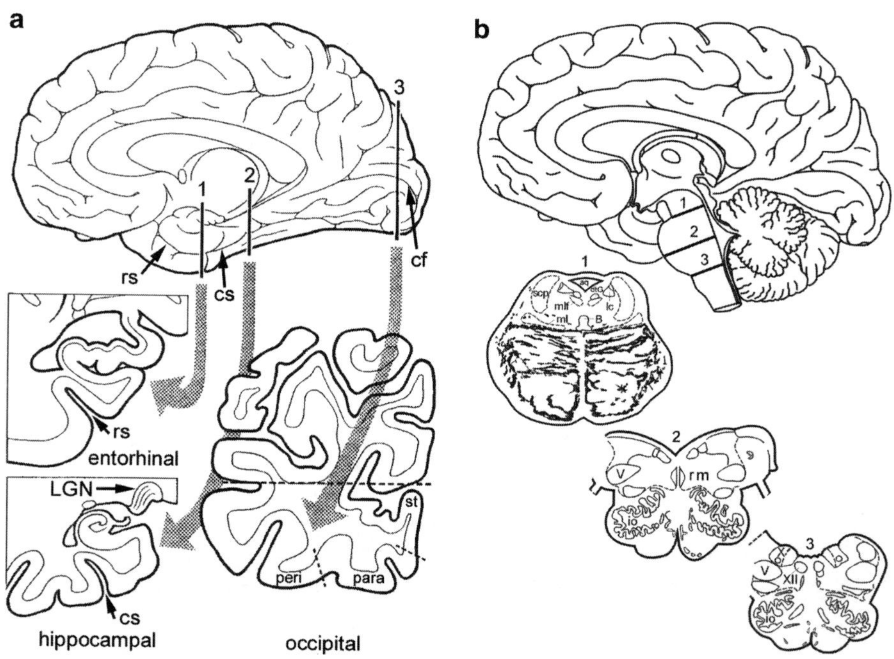

Fig. 11.1 Tissue blocks recommended for neuropathological staging of AD. Abbreviations: **aq**—mesencephalic aqueduct; **cf**—calcarine fissure; **cs**—collateral sulcus; **dtG**—dorsal tegmental nucleus of Gudden; **io**—inferior olivary complex; **lc**—locus coeruleus; **LGN**—lateral geniculate nucleus; **ml**—medial lemniscus; ***mlf***—medial longitudinal fascicle; **para**—parastriate area; **peri**—peristriate area; **rm**—great raphe nucleus; **rs**—rhinal sulcus; ***scp***—decussation of superior cerebellar peduncle; **st**—striate area; **V**—motor nucleus of the trigeminal nerve; **X**—dorsal motor nucleus of the vagal nerve; **XII**—motor nucleus of the hypoglossal nerve (Adapted with permission from H Braak and E Braak, Exp Neurol 1998;153:351–365)

Bodian). This may be advantageous in some instances but usually impedes differentiation of the changed components from uninvolved parts of the brain.

Fortunately, the conventional silver methods can be replaced by modern techniques, which take advantage of physical development of the nucleation sites and thereby permit tight control of the entire procedure. These techniques are much simpler than the conventional methods and far more reliable. The result is almost identical to that obtained by immunoreactions, and the lesions are made visible with virtually no distracting background staining. The techniques can be applied to routinely fixed autopsy material, even if the material has been stored for decades in formalin solutions. They facilitate the processing of large numbers and/or large-sized sections (double hemisphere sections). Both thin paraffin sections (5–15 μm) and thick frozen or PEG sections (50–150 μm) can be processed. For easy identification of architectonic units it is possible to counterstain paraffin sections for Nissl material and/or other structures. Contrary to the situation in immunostained preparations, a homogeneous staining throughout the entire thickness of a section is achieved, this frequently being a prerequisite for quantitative evaluations. The methods are inexpensive and can be applied almost anywhere in the world.

11.1 Stock Solution for Physical Developer

Prepare stock solution I by dissolving 50 g anhydrous sodium carbonate (Merck 6392) in 1,000 ml deionized water. Prepare stock solution II by consecutively dissolving 2 g ammonium nitrate (Merck 1188), 2 g silver nitrate, 10 g tungstosilicic acid ($H_4[Si(W_3O_{10})_4] \times H_2O$, Merck 659) in 1,000 ml deionized water. Prepare stock solution III by consecutively dissolving 2 g ammonium nitrate, 2 g silver nitrate, 10 g tungstosilicic acid, and 7.3 ml of a 35 % aqueous solution of formaldehyde in 1,000 ml distilled water. Stock solutions II and III should always be stored in dark bottles.

11.2 Campbell-Switzer Technique for Brain-Amyloid Deposits

Induce silver nucleation sites by placing sections in a pyridine silver solution for 40 min. Use a chemical hood for preparing the solution by adding 2 volume units pyridine (Merck 822301) to 12 volume units of a 1 % aqueous solution of silver nitrate. Add 9 volume units of a 1 % aqueous solution of potassium carbonate ($K_2CO_3.2H_2O$). Wash in a 0.5 % aqueous solution of acetic acid and in deionized water, each for 3 min.

Visualize nucleation sites by physical development for 5–30 min. Evaluation of the time needed for sufficient staining is facilitated by co-development of a control-section with abundant amyloid deposits. Prepare developer fresh by adding 9 volume units of stock solution II to 10 volume units stock solution I, stir and add 1 volume unit stock solution III, stir and wait until clear. Wash in 0.5 % acetic acid and in deionized water, each for 3 min.

Fix the reaction product in a 1 % aqueous solution of sodium thiosulfate ($Na_2S_2O_3.5H_2O$) for 5 min. Wash in deionized water for 5 min. Stabilize and tone sections in a 0.2 % aqueous solution of yellow gold chloride ($HAuCl_4 \times 4H_2O$, Merck 1582) for 5 min. Fix again in a 1 % aqueous solution of $Na_2S_2O_3$. Wash in deionized water for 5 min.

If desired, counterstain paraffin sections using a suitable Nissl stain (cresyl violet or Darrow red). Dehydrate sections, clear, and mount in a synthetic resin (Histomount, National Diagnostics). To obtain flatness, place free floating PEG sections between sheets of blotting paper and compress gently with porcelain desiccator plates during dehydration in 70 %, 96 % (each for 30 min), and 100 % ethanol (30 min). Free sections and dehydrate for another 30 min in 100 % ethanol. Clear in two portions of xylene, each for 30 min, coverslip, and mount in a synthetic resin (Histomount, National Diagnostics).

Brain-amyloid and congophilic angiopathy appear in a deep purple shade with high contrast against an unstained background. Other amyloid precipitations such as those occurring in spongiform encephalopathies remain unstained. Fully developed sections show a non-specific co-staining of axons. The technique also distinctly demonstrates neuromelanin granules, as well as Lewy bodies/neurites related to Lewy body disease (Parkinson's disease) and argyrophilic oligodendrocytes associated with multisystem atrophy.

11.3 Gallyas Technique for Neurofibrillary Pathology

Pre-treat thick free-floating frozen or PEG sections in a 5 % aqueous solution of periodic acid (HJO_4, Merck Schuchardt 822288; can be used up to six times) for 10 min. Wash repeatedly in deionized water for 10 min. Pre-treat paraffin sections in lanthanum nitrate for 1 h. Prepare solution by dissolving 0.5 g lanthanum nitrate ($La(NO_3)_3 \times 6H_2O$, Merck 5326) and 2 g sodium acetate ($CH_3COOHNa \times 3H_2O$, Merck 6267) in 90 ml deionized water and add 10 ml perhydrol (30 % H_2O_2, hydrogenperoxide, Merck 1.07209). Wash in deionized water for 1 min.

Induce nucleation sites by transferring sections into an alkaline solution of silver iodide (1 min for paraffin material to prevent detachment of sections from the slides; 10 min for free-floating thick sections). Prepare solution by dissolving 40 g sodium hydroxide (NaOH, Merck 6498) in 500 ml distilled water, add 100 g potassium iodide (KJ, Merck 5040), stir and wait until dissolved. Add slowly 35 ml of a 1 % aqueous solution of silver nitrate ($AgNO_3$) stir until clear, then add distilled water to achieve a final volume of 1,000 ml. Rinse sections in deionized water for about 10 min.

Visualize neurofibrillary changes by physical development for 5–30 min. Evaluation of the time needed is facilitated by co-development of a control section with abundant changes. This is necessary when staining brain tissue with little changes escaping recognition when controlled with the unaided eye. Prepare fresh developer and adjust to room temperature by adding suitable volume units of stock solution II to those of stock solution I. Stir and add a sufficient volume of stock solution III, stir, and wait until clear.

Room temperature	I [ml]	II [ml]	III [ml]
20 °C	50	15	35
25 °C	50	25	25
30 °C	50	35	15
35 °C	50	40	10

Rinse in 0.5 % acetic acid and in deionized water for 3 min. Stabilize and mount sections as described above. If desired, counterstain sections.

Neurofibrillary changes of the Alzheimer-type (NFTs, NTs, NPs) are stained in a black shade and contrast well against the unstained background. Abnormal tau-protein in nerve cells and/or macroglial cells in argyrophilic grain disease (AGD), PSP, CBD, and Niemann-Pick disease, type C (NP-C) are also shown. Connective tissue, glial filaments, normal components of the neuronal cytoskeleton, as well as Pick bodies, Lewy bodies/neurites, corpora amylacea, etc. remain unstained.

References

Agnati LF, Bjelke B, Fuxe K (1995) Volume versus wiring transmission in the brain: a new theoretical frame of neuropsychopharmacology. Med Res Rev 15:33–45

Ahmed Z, Cooper J, Murray TK et al (2014) A novel *in vivo* model of tau propagation with rapid and progressive neurofibrillary tangle pathology: the pattern of spread is determined by connectivity, not proximity. Acta Neuropathol 127:667–683

Alafuzoff I, Pikkairainen M, Al-Sarraj S et al (2006) Staging of Alzheimer disease-associated neurofibrillary pathology using paraffin sections and immunocytochemistry. J Neuropathol Exp Neurol 65:740–757

Alafuzoff I, Thal DR, Arzberger T et al (2009) Assessment of beta-amyloid deposits in human brain: a study of the BrainNet Europe Consortium. Acta Neuropathol 117:309–320

Albert MS, Dekosky ST, Dickson D et al (2011) The diagnosis of mild cognitive impairment due to Alzheimer's disease: recommendations from the National Institute on Aging-Alzheimer's association workgroups on diagnostic guidelines for Alzheimer's disease. Alzheimers Dement 7:270–279

Alheid GF (2003) Extended amygdala and basal forebrain. Ann NY Acad Sci 985:185–205

Alonso AC, Li B, Grundke-Iqbal I, Iqbal K (2008) Mechanism of tau-induced neurodegeneration in Alzheimer disease and related tauopathies. Curr Alzheimer Res 5:375–384

Alzheimer A (1906) Über einen eigenartigen schweren Erkrankungsprozeß der Hirnrinde. Neurolog Centralbl 23:1129–1136

Alzheimer A, Stelzmann RA, Schnitzlein HN, Murtagh R (1995) An English translation of Alzheimer's 1907 paper, „Über eine eigenartige Erkrankung der Hirnrinde". Clin Anat 8:429–431

Amieva H, Le Goff M, Millet X et al (2008) Prodromal Alzheimer's disease: successive emergence of clinical symptoms. Ann Neurol 64:492–498

Amunts K, Zilles K (2001) Advances in cytoarchitectonic mapping of the human cerebral cortex. Neuroimaging Clin N Am 11:151–169

Amunts K, Kedo O, Kindler M et al (2005) Cytoarchitectonic mapping of the human amygdala, hippocampal region and entorhinal cortex: intersubject variability and probability maps. Anat Embryol 210:343–352

Andreasen N, Minthon L, Vanmechelen E et al (1999) Cerebrospinal fluid tau and Abeta42 as predictors of development of Alzheimer's disease in patients with mild cognitive impairment. Neurosci Lett 273:5–8

Arendt T (2000) Alzheimer's disease as a loss of differentiation control in a subset of neurons that retain immature features in the adult brain. Neurobiol Aging 21:783–796

H. Braak, K. Del Tredici, *Neuroanatomy and Pathology of Sporadic Alzheimer's Disease*, Advances in Anatomy, Embryology and Cell Biology 215,
DOI 10.1007/978-3-319-12679-1

Arendt T (2004) Neurodegeneration and plasticity. Int J Dev Neurosci 22:507–514

Arendt T (2005) Alzheimer's disease as a disorder of dynamic brain self-organization. Prog Brain Res 147:355–378

Arendt T (2012) Cell cycle activation and aneuploid neurons in Alzheimer's disease. Mol Neurobiol 46:125–135

Arendt T, Brückner MK, Gertz HJ, Marcova L (1998) Cortical distribution of neurofibrillary tangles in Alzheimer's disease matches the pattern of neurones that retain their capacity of plastic remodelling in the adult brain. Neuroscience 83:991–1002

Arnold SE, Hyman BT, Flory J et al (1991) The topographical and neuroanatomical distribution of neurofibrillary tangles and neuritic plaques in the cerebral cortex of patients with Alzheimer's disease. Cereb Cortex 1:103–116

Arnold SE, Lee EB, Moberg PJ et al (2010) Olfactory epithelium amyloid-β and PHFtau pathology in Alzheimer's disease. Ann Neurol 67:462–469

Arshavsky YI (2014) Alzheimer disease and cellular mechanisms of memory storage. J Neuropathol Exp Neurol 73:192–205

Ashford JW, Bayley PJ (2013) Retrogenesis: a model of dementia progression in Alzheimer's disease related to neuroplasticity. J Alzheimers Dis 33:1191–1193

Ashford JW, Soultanian NS, Zhang SX, Geddes JW (1998) Neuropil threads are collinear with MAP2 immunostaining in neuronal dendrites of Alzheimer brain. J Neuropathol Exp Neurol 57:972–978

Aston-Jones C, Cohen JD (2005) An integrative theory of locus coeruleus-norepinephrine function: adaptive gain and optimal performance. Ann Rev Neurosci 28:403–450

Attems J, Jellinger KA (2006) Olfactory tau pathology in Alzheimer disease and mild cognitive impairment. Clin Neuropathol 25:265–271

Attems J, Jellinger K, Thal DR, Van Nostrand W (2011) Review: sporadic cerebral amyloid angiopathy. Neuropathol Appl Neurobiol 37:75–93

Attems J, Jellinger K, Thal DR (2012) The relationship between subcortical tau pathology and Alzheimer's disease. Biochem Soc Trans 40:711–715

Augustinack JC, Schneider A, Mandelkow EM, Hyman BT (2002) Specific tau phosphorylation sites correlate with severity of neuronal cytopathology in Alzheimer's disease. Acta Neuropathol 103:26–35

Augustinack JC, Huber KE, Postelnicu GM et al (2012) Entorhinal verrucae geometry is coincident and correlates with Alzheimer's lesions: a combined neuropathology and high-resolution *ex vivo* MRI analysis. Acta Neuropathol 123:85–96

Ávila J (2006) Tau phosphorylation and aggregation in Alzheimer's disease pathology. FEBS Lett 580:2922–2927

Ballatore C, Lee VMY, Trojanowski JQ (2007) Tau-mediated neurodegeneration in Alzheimer's disease and related disorders. Nat Rev Neurosci 8:663–672

Bancher C, Brunner C, Lassmann H et al (1989) Accumulation of abnormally phosphorylated tau precedes the formation of neurofibrillary tangles in Alzheimer's disease. Brain Res 477:90–99

Bannister AP (2005) Inter- and intra-laminar connections of pyramidal cells in the neocortex. Neurosci Res 53:95–103

Barbas H (2007) Specialized elements of orbitofrontal cortex in primates. Ann NY Acad Sci 1121:10–32

Bartzokis G (2004) Age-related myelin breakdown: a developmental model of cognitive decline and Alzheimer's disease. Neurobiol Aging 25:5–18

Barulli D, Stern Y (2013) Efficiency, capacity, compensation, maintenance, plasticity: emerging concepts in cognitive reserve. Trends Cogn Sci 17:502–509

Beach TG, Monsell SE, Phillips LE, Kukull W (2012a) Accuracy of the clinical diagnosis of Alzheimer's disease at National Institute on Aging Alzheimer Disease Centers, 2005–2010. J Neuropathol Exp Neurol 71:266–273

Beach TG, Sue LI, Walker DG et al (2012b) Striatal amyloid plaque density predicts Braak neurofibrillary stage and clinicopathological Alzheimer's disease: implications for amyloid imaging. J Alzheimers Dis 28:869–876
Benarroch EE (1993) The central autonomic network: functional organization, dysfunction, and perspective. Mayo Clin Proc 68:988–1001
Benarroch EE (2009) The locus ceruleus norepinephrine system. Functional organization and potential clinical significance. Neurology 17:1699–1704
Benarroch EE (2013) Neocortical interneurons. Functional diversity and clinical correlations. Am Acad Neurol 81:273–280
Bengtsson SL, Nagy Z, Skare S et al (2005) Extensive piano practicing has regionally specific effects on white matter development. Nat Neurosci 8:1148–1150
Beyreuther K, Masters CL (1991) Amyloid precursor protein (APP) and beta amyloid-42 amyloid in the etiology of Alzheimer's disease: precursor product relationships in the derangement of neuronal function. Brain Pathol 1:241–252
Binder LI, Guillozet-Bongaarts AL, Garcia-Sierra F, Berry RW (2004) Tau, tangles, and Alzheimer's disease. Biochim Biophys Acta 1739:216–223
Blazquez-Llorca L, Garcia-Martin V, Merino-Serrais P et al (2011) Abnormal tau phosphorylation in the thorny excrescences of CA3 hippocampal neurons in patients with Alzheimer's disease. J Alzheimer Dis 23:1–16
Blennow K, Hampel H (2003) Cerebrospinal fluid markers for incipient Alzheimer's disease. Lancet Neurol 2:605–613
Blennow K, Zetterberg H, Minthon L et al (2007) Longitudinal stability of CSF biomarkers in Alzheimer's disease. Neurosci Lett 419:18–22
Blennow K, Hampel H, Weiner M, Zetterberg H (2010) Cerebrospinal fluid and plasma biomarkers in Alzheimer disease. Nat Rev Neurol 6:131–144
Blennow K, Zetterberg H, Fagan AM (2012) Fluid biomarkers in Alzheimer's disease. Cold Spring Harb Perspect Med 2:a006221
Blessing WW (2004) Lower brain stem regulation of visceral, cardiovascular, and respiratory function. In: Paxinos G, Mai JK (eds) The human nervous system, 2nd edn. Elsevier, San Diego, pp 464–478
Blom ES, Giedraitis V, Zetterberg H et al (2009) Rapid progression from mild cognitive impairment to Alzheimer's disease in subjects with elevated levels of tau in cerebrospinal fluid and the APOE epsilon4/epsilon4 genotype. Dement Geriatr Cogn Disord 27:458–464
Bobinski M, Wegiel J, Tarnawski M et al (1998) Duration of neurofibrillary changes in the hippocampal pyramidal neurons. Brain Res 799:156–158
Bohus B, Koolhaas JM, Luiten PGM et al (1996) The neurobiology of the central nuleus of the amygdala in relation to neuroendocrine and autonomic outflow. Prog Brain Res 107:447–460
Bonthius DJ, Solodkin A, van Hoesen GW (2005) Pathology of the insular cortex in Alzheimer disease depends on cortical architecture. J Neuropathol Exp Neurol 64:910–922
Bouras C, Hof P, Giannakopoulos P et al (1994) Regional distribution of neurofibrillary tangles and senile plaques in the cerebral cortex of elderly patients: a quantitative evaluation of a one-year autopsy population from a geriatric hospital. Cereb Cortex 4:138–150
Braak H (1979) Spindle-shaped appendages of IIIab-pyramids filled with lipofuscin: a striking pathological change of the senescent human isocortex. Acta Neuropathol 46:197–202
Braak H (1980) Architectonics of the human telencephalic cortex. Springer, Berlin, pp 1–147
Braak H, Braak E (1990) Alzheimer's disease: amyloid deposits and neurofibrillary changes in the striatum. J Neuropathol Exp Neurol 49:215–224
Braak H, Braak E (1991a) Neuropathological stageing of Alzheimer-related changes. Acta Neuropathol 82:239–259
Braak H, Braak E (1991b) Demonstration of amyloid deposits and neurofibrillary changes in whole brain sections. Brain Pathol 1:213–216
Braak H, Braak E (1992a) The human entorhinal cortex: normal morphology and lamina-specific pathology in various diseases. Neurosci Res 15:6–31

Braak H, Braak E (1992b) Anatomy of the human hypothalamus (chiasmatic and tuberal region). Prog Brain Res 93:3–16

Braak H, Braak E (1995) Staging of Alzheimer's disease-related neurofibrillary changes. Neurobiol Aging 16:271–284

Braak H, Braak E (1996) Development of Alzheimer-related neurofibrillary changes in the neocortex inversely recapitulates cortical myelogenesis. Acta Neuropathol 92:197–201

Braak E, Braak H (1997a) Alzheimer's disease: transiently developing dendritic changes in pyramidal cells of sector CA1 of the Ammon's horn. Acta Neuropathol 93:323–325

Braak H, Braak E (1997b) Frequency of stages of Alzheimer-related lesions in different age categories. Neurobiol Aging 18:351–357

Braak H, Braak E (1999) Temporal sequence of Alzheimer's disease-related pathology. In: Peters A, Morrison JH (eds) Cerebral cortex, vol 14, Neurodegenerative and age-related changes in structure and function of the cerebral cortex. Kluwer Academic/Plenum, New York, NY, pp 475–512

Braak H, Del Tredici K (2004) Poor and protracted myelination as a contributory factor to neurodegenerative disorders. Neurobiol Aging 25:19–23

Braak H, Del Tredici K (2009) Neuroanatomy and pathology of sporadic Parkinson's disease. Adv Anat Embryol Cell Biol 201:1–119

Braak H, Del Tredici K (2011) The pathological process underlying Alzheimer's disease in individuals under thirty. Acta Neuropathol 121:171–181

Braak H, Del Tredici K (2012) Where, when, and in what form does sporadic Alzheimer's disease begin? Curr Opin Neurol 25:708–714

Braak H, Del Tredici K (2013a) Amyloid-β may be released from non-junctional varicosities of axons generated from abnormal tau-containing brainstem nuclei in sporadic Alzheimer's disease: a hypothesis. Acta Neuropathol 126:303–306

Braak H, Del Tredici K (2013b) Reply: the early pathological process in sporadic Alzheimer's disease. Acta Neuropathol 126:615–681

Braak H, Del Tredici K (2014) Are cases with tau pathology occurring in the absence of the Aβ deposits part of the AD-related pathological process? Acta Neuropathol 128:767–772

Braak H, Braak E, Kalus P (1989a) Alzheimer's disease: areal and laminar pathology in the occipital isocortex. Acta Neuropathol 77:494–506

Braak H, Braak E, Bohl J, Lang W (1989b) Alzheimer's disease: amyloid plaques in the cerebellum. J Neurol Sci 93:277–287

Braak E, Braak H, Mandelkow EM (1994) A sequence of cytoskeleton changes related to the formation of neurofibrillary tangles and neuropil threads. Acta Neuropathol 87:554–567

Braak H, Braak E, Yilmazer D, Bohl J (1996) Functional anatomy of human hippocampal formation and related structures. J Child Neurol 11:265–275

Braak H, Del Tredici K, Braak E (2003) Spectrum of pathology. In: Petersen RC (ed) Mild cognitive impairment. Aging to Alzheimer's disease. Oxford University Press, Oxford, pp 149–189

Braak H, Alafuzoff I, Arzberger T et al (2006a) Staging of Alzheimer's disease-associated neurofibrillary pathology using paraffin sections and immunohistochemistry. Acta Neuropathol 112:389–404

Braak H, Rüb U, Schultz C, Del Tredici K (2006b) Vulnerability of cortical neurons to Alzheimer's and Parkinson's diseases. J Alzheimers Dis 9(3 Suppl):35–44

Braak H, Thal DR, Ghebremedhin E, Del Tredici K (2011) Stages of the pathological process in Alzheimer disease: age categories from 1 to 100 years. J Neuropathol Exp Neurol 70:960–969

Braak H, Zetterberg H, Del Tredici K, Blennow K (2013) Intraneuronal tau aggregation precedes diffuse plaque deposition, but amyloid-β changes occur before increases of tau in cerebrospinal fluid. Acta Neuropathol 126:631–641

Brookmeyer R, Johnson E, Ziegler-Graham K, Arrighi HM (2007) Forecasting the global burden of Alzheimer's disease. Alzheimers Dement 3:186–191

Brunden KR, Trojanowski JQ, Lee VMY (2008) Evidence that non-fibrillar tau causes pathology linked to neurodegeneration and behavioral impairments. J Alzheimers Dis 14:393–399

Buchhave P, Minthon L, Zetterberg H et al (2012) Cerebrospinal fluid levels of β-amyloid 1-42, but not of tau, are fully changed already 5 to 10 years before the onset of Alzheimer dementia. Arch Gen Psychiatry 69:98–106

Buée L, Bussiere T, Buée-Scherrer V et al (2000) Tau protein isoforms, phosphorylation and role in neurodegenerative disorders. Brain Res Rev 33:95–130

Buerger K, Ewers M, Pirttilä T et al (2006) CSF phosphorylated tau protein correlates with neocortical neurofibrillary pathology in Alzheimer's disease. Brain 129:3035–3041

Bufill E, Blesa R, August J (2013) Alzheimer's disease: an evolutionary approach. J Anthropol Sci 91:135–157

Bulic B, Pickhardt M, Mandelkow EM, Mandelkow E (2010) Tau protein and tau aggregation inhibitors. Neuropharmacology 59:276–289

Busch C, Bohl J, Ohm TG (1997) Spatial, temporal and numeric analysis of Alzheimer changes in the nucleus coeruleus. Neurobiol Aging 18:401–406

Butti C, Santos M, Uppal N, Hof PR (2013) Von Economo neurons: clinical and evolutionary perspectives. Cortex 49:312–326

Buxbaum JD, Thinakaran G, Koliatsos V et al (1998) Alzheimer amyloid protein precursor in the rat hippocampus: transport and processing through the perforant path. J Neurosci 18:9629–9637

Bzdok D, Laird AR, Zilles K et al (2013) An investigation of the structural, connectional, and functional subspecialization in the human amygdala. Hum Brain Mapp 34:3247–3266

Campbell SK, Switzer RC, Martin TL (1987) Alzheimer's plaques and tangles: a controlled and enhanced silver-staining method. Soc Neurosci Abstr 13:678

Chalermpalanupap T, Kinkead B, Hu WT et al (2013) Targeting norepinephrine in mild cognitive impairment and Alzheimer's disease. Alzheimers Res Ther 5:21

Chamberlain SR, Robbins TW (2013) Noradrenergic modulation of cognition: therapeutic implications. J Psychopharmacol 8:694–718

Chan-Palay V, Asan E (1989) Alterations in catecholamine neurons of the locus coeruleus in senile dementia of the Alzheimer type and in Parkinson's disease with and without dementia and depression. J Comp Neurol 287:373–392

Chételat G (2013) Alzheimer disease: Aβ-independent processes-rethinking preclinical AD. Nat Rev Neurol 9:123–124

Chételat G, Fouquet M (2013) Neuroimaging biomarkers for Alzheimer's disease in asymptomatic APOE4 carriers. Rev Neurol 169:729–736

Christen-Zaech S, Kraftsik R, Pillevuit O et al (2003) Early olfactory involvement in Alzheimer's disease. Can J Neurol Sci 30:20–25

Chu CC, Tranel D, Damasio AR et al (1997) The autonomic-related cortex: pathology in Alzheimer's disease. Cereb Cortex 7:86–95

Chui HC, Zheng L, Reed BR et al (2012) Vascular risk factors and Alzheimer's disease: are these risk factors for plaques and tangles or for concomitant vascular pathology that increases the likelihood of dementia? An evidence-based review. Alzheimers Res Ther 4:1

Clavaguera F, Bolmont T, Crowther RA et al (2009) Transmission and spreading of tauopathy in transgenic mouse brain. Nat Cell Biol 11:909–913

Clavaguera F, Akatsu H, Fraser G et al (2013a) Brain homogenates from human tauopathies induce tau inclusions in mouse brain. Proc Natl Acad Sci USA 110:9535–9540

Clavaguera F, Lavenir I, Falcon B et al (2013b) "Prion-like" templated misfolding in tauopathies. Brain Pathol 23:342–349

Clavaguera F, Grueninger F, Tolnay M (2014) Intercellular transfer of tau aggregates and spreading of tau pathology: implications for therapeutic strategies. Neuropharmacology 76 (Pt A):9–15

Counts SE, Mufson EJ (2012) Locus coeruleus. In: Mai JK, Paxinos G (eds) The human nervous system, 3rd edn. Academic Press, New York, NY, pp 425–438

Cower CM, Mudher A (2013) Are tau aggregates toxic or protective in tauopathies? Front Neurol 4:114–126
Craig-Schapiro R, Fagan AM, Holtzman DM (2009) Biomarkers of Alzheimer's disease. Neurobiol Dis 35:128–140
Craik FIM, Bialystok E, Freedman M (2010) Delaying the onset of Alzheimer's disease. Bilingualism as a form of cognitive reserve. Neurology 75:1726–1729
Cramer SC, Chopp M (2000) Recovery recapitulates ontogeny. Trends Neurosci 23:265–271
Crary JF, Trojanowski JQ, Schneider JA et al (2014) Primary age-related tauopathy (PART): a common pathology associated with aging. Acta Neuropathol 128:755–766
Davis DG, Schmitt FA, Wekstein DR, Markesbery WR (1999) Alzheimer neuropathologic alterations in aged cognitively normal subjects. J Neuropathol Exp Neurol 58:376–388
de Calignon A, Polydoro M, Suarez-Calvet M et al (2012) Propagation of tau pathology in a model of early Alzheimer's disease. Neuron 73:685–697
de la Monte SM, Tong M (2013) Brain metabolic dysfunction at the core of Alzheimer's disease. Biochem Pharmacol 88:548–559
de Silva R, Lashley T, Gibb G et al (2003) Pathological inclusion bodies in tauopathies contain distinct complements of tau with three or four microtubule-binding repeat domains as demonstrated by new specific monoclonal antibodies. Neuropathol Appl Neurobiol 29:288–302
DeFelipe J, Alonso-Nanclares L, Arellano JI (2002) Microstructure of the neocortex: comparative aspects. J Neurocytol 31:299–316
Del Tredici K, Braak H (2012) Spinal cord lesions in sporadic Parkinson's disease. Acta Neuropathol 124:643–664
Del Tredici K, Hawkes CH, Ghebremedhin E, Braak H (2010) Lewy pathology in the submandibular gland of individuals with incidental Lewy body disease and sporadic Parkinson's disease. Acta Neuropathol 119:703–713
Delacourte A (2008) Tau pathology and neurodegeneration: an obvious but misunderstood link. J Alzheimers Dis 14:437–440
Delacourte A, David JP, Sergeant N et al (1999) The biochemical pathway of neurofibrillary degeneration in aging and Alzheimer's disease. Neurology 52:1158–1165
Delbeuck X, van der Linden M, Collette F (2003) Alzheimer's disease as a disconnection syndrome? Neuropsychol Rev 13:79–92
DeLong MR, Wichmann T (2007) Circuits and circuit disorders of the basal ganglia. Arch Neurol 64:20–24
Dickerson BC, Stoub TR, Shah RC et al (2011) Alzheimer-signature MRI biomarker predicts AD dementia in cognitively normal adults. Neurology 76:1395–1402
Dickson DW (1997a) The value of cross-sectional neuroanatomical studies as a conceptual framework for prospective clinicopathological studies. Neurobiol Aging 18:382–386
Dickson DW (1997b) The pathogenesis of senile plaques. J Neuropathol Exp Neurol 56:321–339
Dickson DW, Vickers JC (2001) The morphological phenotype of beta-amyloid plaques and associated neuritic changes in Alzheimer's disease. Neuroscience 105:99–107
Dickson DW, Rademakers R, Hutton ML (2007) Progressive supranuclear palsy: pathology and genetics. Brain Pathol 17:74–82
Ding H, Johnson GV (2008) The last tangle of tau. J Alzheimers Dis 14:441–447
Ding SL, van Hoesen GW (2010) Borders, extent, and topography of human perirhinal cortex as revealed using multiple modern neuroanatomical and pathological markers. Hum Brain Mapp 31:1359–1379
Dolan D, Troncoso J, Resnick SM et al (2010) Age, Alzheimer's disease and dementia in the Baltimore longitudinal study of ageing. Brain 133:2225–2231
Dong S, Duan Y, Hu Y, Zhao Z (2012) Advances in the pathogenesis of Alzheimer's disease: a re-evaluation of amyloid cascade hypothesis. Transl Neurodegener 1:1–18
Doody RS, Thomas RG, Farlow M et al (2014) Phase 3 trials of solanezumab for mild-to-moderate Alzheimer's disease. N Engl J Med 370:311–321

Double KL, Dedov VN, Fedorow H et al (2008) The comparative biology of neuromelanin and lipofuscin in the human brain. Cell Mol Life Sci 11:1669–1682
Draganski B, May A (2008) Training-induced structural changes in the adult human brain. Behav Brain Res 192:137–142
Dringenberg HC (2000) Alzheimer's disease: more than a 'cholinergic disorder' – evidence that cholinergic-monoaminergic interactions contribute to EEG slowing and dementia. Behav Brain Res 115:236–249
Dubois B, Feldman HH, Jacova C et al (2010) Revising the definition of Alzheimer's disease: a new lexicon. Lancet Neurol 9:1118–1127
Dugger BN, Tu M, Murray ME, Dickson DW (2011) Disease specificity and pathologic progression of tau pathology in brainstem nuclei of Alzheimer's disease and progressive supranuclear palsy. Neurosci Lett 491:122–126
Dugger BN, Hidalgo JA, Chiarolanza G et al (2013) The distribution of phosphorylated tau in spinal cords of Alzheimer's disease and non-demented individuals. J Alzheimer Dis 34:529–536
Dujardin S, Lécolle K, Caillierez R, Bégard S et al (2014) Neuron-to-neuron wild-type tau protein transfer through a trans-synaptic mechanism: relevance to sporadic tauopathies. Acta Neuropathol Commun 2:14
Duyckaerts C (2011) Tau pathology in children and young adults: can you still be unconditionally baptist? Acta Neuropathol 121:145–147
Duyckaerts C (2013) Neurodegenerative lesions: seeding and spreading. Rev Neurol 169:825–833
Duyckaerts C, Hauw JJ (1997) Prevalence, incidence and duration of Braak's stages in the general population: can we know? Neurobiol Aging 18:362–369
Duyckaerts C, Delatour B, Potier MC (2009) Classification and basic pathology of Alzheimer's disease. Acta Neuropathol 118:5–36
Elfenbein HA, Rosen RF, Stephens SL et al (2007) Cerebral beta-amyloid angiopathy in aged squirrel monkeys. Histol Histopathol 22:155–167
Elobeid A, Soininen H, Alafuzoff I (2011) Hyperphosphorylated tau in young and middle-aged subjects. Acta Neuropathol 123:97–104
Englund H, Degerman Gunnarsson M, Brundin RM et al (2009) Oligomerization partially explains the lowering of Aβ42 in Alzheimer's disease cerebrospinal fluid. Neurodegener Dis 6:139–147
Erickson KI, Weinstein AM, Lopez OL (2012) Physical activity, brain plasticity, and Alzheimer's disease. Arch Med Res 43:615–621
España RA, Berridge CW (2006) Organization of noradrenergic efferents to arousal-related basal forebrain. J Comp Neurol 496:668–683
Ewers M, Frisoni GB, Teipel SJ et al (2011) Staging Alzheimer's disease progression with multimodality neuroimaging. Prog Neurobiol 95:535–546
Ewers M, Walsh C, Trojanowski JQ et al (2012) Prediction of conversion from mild cognitive imairment to Alzheimer's disease dementia based upon biomarkers and neuropsychological test performance. Neurobiol Aging 33:1203–1214
Fagan AM, Mintun MA, Shah AR et al (2009) Cerebrospinal fluid tau and ptau(181) increase with cortical amyloid deposition in cognitively normal individuals: implications for future clinical trials of Alzheimer's disease. EMBO Mol Med 1:371–380
Farkas E, Luiten PGM (2001) Cerebral microvascular pathology in aging and Alzheimer's disease. Prog Neurobiol 64:575–611
Feldengut S, Del Tredici K, Braak H (2013) Paraffin sections of 70-100 μm: a novel technique and its benefits for studying the nervous system. J Neurosci Method 215:241–244
Femminella GD, Rengo G, Komici K et al (2014) Autonomic dysfunction in Alzheimer's disease: tools for assessment and review of the literature. J Alzheimers Dis 42:369–377
Ferrer I (2012) Defining Alzheimer as a common age-related neurodegenerative process not inevitably leading to dementia. Prog Neurobiol 97:38–51
Fiala JC (2007) Mechanisms of amyloid plaque pathogenesis. Acta Neuropathol 114:551–571
Fields RD (2011) Imaging learning: the search for a memory trace. Neuroscientist 17:185–196

Fodero-Tavoletti MT, Okamura N, Furumoto S et al (2011) 18F-THK523: a novel in vivo tau imaging ligand for Alzheimer's disease. Brain 134:1089–1100

Fodero-Tavoletti MT, Furumoto S, Taylor L et al (2014) Assessing THK523 selectivity for tau deposits in Alzheimer's disease and non-Alzheimer's disease tauopathies. Alzheimers Res Ther 6:11–21

Fotuhi M, Hachinski V, Whitehouse PJ (2009) Changing perspectives regarding late-life dementia. Nat Rev Neurol 5:649–658

Frackowiac J, Zoltowska A, Wisniewski HM (1994) Non-fibrillar beta-amyloid protein is associated with smooth muscle cells of vessel walls in Alzheimer disease. J Neuropathol Exp Neurol 53:637–645

Frankfort SV, Tulner LR, van Campen JP et al (2008) Amyloid beta protein and tau in cerebrospinal fluid and plasma as biomarkers for dementia: a review of recent literature. Curr Clin Pharmacol 3:123–131

Franssen EH, Kluger A, Torossian CL, Reisberg B (1993) The neurologic syndrome of severe Alzheimer's disease. Relationship to functional decline. Arch Neurol 50:1029–1039

Freedman M, Alladi S, Chertkow H et al (2014) Delaying onset of dementia: are two languages enough? Behav Neurol 2014, 808137

Friedman JI, Adler DH, Davis KL (1999) The role of norepinephrine in the pathophysiology of cognitive disorders: potential applications to the treatment of cognitive dysfunction in Schizophrenia and Alzheimer's disease. Biol Psychiatry 46:1243–1252

Frost B, Diamond MI (2010) Prion-like mechanisms in neurodegenerative diseases. Nat Rev Neurosci 11:155–159

Gallyas F (1971) Silver staining of Alzheimer's neurofibrillary changes by means of physical development. Acta Morphol Acad Sci Hung 19:1–8

Garcia-Sierra F, Ghoshal N, Quinn B et al (2003) Conformational changes and truncation of tau protein during tangle evolution in Alzheimer's disease. J Alzheimer's disease 5:65–77

Gärtner H, Minnerop M, Pieperhoff P et al (2013) Brain morphometry shows effects of long-term musical practice in middle-aged keyboard players. Front Psychol 4:636

Gates GA, Beiser A, Rees TS et al (2002) Central auditory dysfunction may precede the onset of clinical dementia in people with probable Alzheimer's disease. J Am Geriatr Soc 50:74–82

German DC, White CL, Sparkman DR (1987) Alzheimer's disease: neurofibrillary tangles in nuclei that project to the cerebral cortex. Neuroscience 21:305–312

German DC, Manaye KF, White CL III et al (1992) Disease-specific patterns of locus coeruleus cell loss. Ann Neurol 32:667–676

Geula C, Mesulam MM (2012) Brainstem cholinergic systems. In: Mai JK, Paxinos G (eds) The human nervous system, 3rd edn. Academic Press, New York, NY, pp 471–547

Ghebremedhin E, Schultz C, Braak E, Braak H (1998) High frequency of apolipoprotein E epsilon4 allele in young individuals with very mild Alzheimer's disease-related neurofibrillary changes. Exp Neurol 153:152–155

Giacobini E, Gold G (2013) Alzheimer disease therapy – moving from amyloid-β to tau. Nat Rev Neurol 9:677–686

Giannakopoulos P, Hof PR, Mottier S et al (1994) Neuropathological changes in the cerebral cortex of 1258 cases from a geriatric hospital: retrospective clinicopathological evaluation of a 10-year autopsy population. Acta Neuropathol 87:456–468

Giannakopoulos P, Herrmann FR, Nussière T et al (2003) Tangle and neuron numbers, but not amyloid load, predict cognitive status in Alzheimer's disease. Neurology 60:1495–1500

Gloor P (1997) The temporal lobe and limbic system. Oxford University Press, New York, NY, pp 1–865

Goedert M, Spillantini MG (2006) A century of Alzheimer's disease. Science 314:777–781

Goedert M, Spillantini MG, Cairns NJ, Crowther RA (1992) Tau proteins from Alzheimer paired helical filaments: abnormal phosphorylation of all six isoforms. Neuron 8:159–168

Goedert M, Trojanowski JQ, Lee VMY (1997) The neurofibrillary pathology of Alzheimer's disease. In: Rosenberg RN (ed) The molecular and genetic basis of neurological disease, 2nd edn. Butterworth-Heinemann, Boston, MA, pp 613–627

Goedert M, Klug A, Crowther RA (2006) Tau protein, the paired helical filament and Alzheimer's disease. J Alzheimers Dis 9(Suppl):195–207

Goedert M, Clavaguera F, Tolnay M (2010) The propagation of prion-like protein inclusions in neurodegenerative diseases. Trends Neurosci 33:317–325

Goedert M, Falcon B, Clavaguera F et al (2014) Prion-like mechanisms in the pathogenesis of tauopathies and synucleinopathies. Curr Neurol Neurosci Rep 14:495

Gogtay N, Thompson PM (2010) Mapping gray matter development: implications for typical development and vulnerability to psychopathology. Brain Cogn 72:6–15

Gold G, Bouras C, Kövari E et al (2000) Clinical validity of Braak neuropathological staging in the oldest-old. Acta Neuropathol 99:579–582

Golde TE, Schneider LS, Koo EH (2011) Anti-aβ therapeutics in Alzheimer's disease: the need for a paradigm shift. Neuron 69:203–213

Gong CX, Grundke-Iqbal I, Iqbal K (2010) Targeting tau protein in Alzheimer's disease. Drugs Aging 27:351–365

Grantham C, Geerts H (2002) The rationale behind cholinergic drug treatment for dementia related to cerebrovascular disease. J Neurol Sci 203–204:131–136

Grimmer T, Henriksen G, Wester HJ et al (2009) Clinical severity of Alzheimer's disease is associated with PiB uptake in PET. Neurobiol Aging 30:1902–1909

Grinberg LT, Thal DR (2010) Vascular pathology in the aged human brain. Acta Neuropathol 119:277–290

Grinberg LT, Rüb U, Ferretti REL et al (2009) The dorsal raphe nucleus shows phospho-tau neurofibrillary changes before the transentorhinal region in Alzheimer's disease. A precocious onset? Neuropathol Appl Neurobiol 35:406–416

Grinberg LT, Korczyn AD, Heinsen H (2012) Cerebral amyloid angiopathy impact on endothelium. Exp Gerontol 47:838–842

Grober E, Dickson D, Sliwinski MJ et al (1999) Memory and mental status correlates of modified Braak staging. Neurobiol Aging 20:573–579

Groenewegen HJ, Trimble M (2007) The ventral striatum as an interface between the limbic and motor systems. CNS Spectr 12:887–892

Grudzien A, Shaw P, Weintraub S et al (2007) Locus coeruleus neurofibrillary degeneration in aging, mild cognitive impairment and early Alzheimer's disease. Neurobiol Aging 28:327–335

Grundke-Iqbal I, Iqbal K, Tung YC et al (1986) Abnormal phosphorylation of the microtubule-associated protein τ (tau) in Alzheimer cytoskeletal pathology. Proc Natl Acad Sci USA 83:4913–4917

Guo JL, Lee VMY (2011) Seeding of normal tau by pathological tau conformers drives pathogenesis of Alzheimer-like tangles. J Biol Chem 286:15317–15331

Guo JL, Lee VMY (2014) Cell-to-cell transmission of pathogenic proteins in neurodegenerative diseases. Nat Med 20:1–9

Haass C, Selkoe DJ (2007) Soluble protein oligomers in neurodegeneration: lessons from the Alzheimer's amyloid β-peptide. Nat Rev Mol Cell Biol 8:101–112

Haass C, Kaether C, Thinakaran G, Sisodia S (2012) Trafficking and proteolytic processing of APP. Cold Spring Harb Perspect Med 2:a006270

Haber SN, Gdowski MJ (2004) The basal ganglia. In: Paxinos G, Mai JK (eds) The human nervous system, 2nd edn. Elsevier, San Diego, CA, pp 677–738

Haglund M, Sjöbeck M, Englund E (2006) Locus ceruleus degeneration is ubiquitous in Alzheimer's disease: possible implications for diagnosis and treatment. Neuropathology 26:528–532

Hall GF, Saman S (2012) Death or secretion? The demise of a plausible assumption about CSF-tau in Alzheimer disease? Commun Integr Biol 5:1–4

Halliday GM, Leverenz JB, Schneider JS, Adler CH (2014) The neurobiological basis of cognitive impairment in Parkinson's disease. Mov Disord 29:634–650
Hamel E (2006) Perivascular nerves and the regulation of cerebrovascular tone. J Appl Physiol 100:1059–1064
Hampel H, Frank R, Broich K et al (2010) Biomarkers for Alzheimer's disease: academic, industry and regulatory perspectives. Nat Rev Drug Discov 9:560–574
Hanke J (1997) Sulcal pattern of the anterior parahippocampal gyrus in the human adult. Ann Anat 179:335–339
Hardy JA (2006) Alzheimer's disease: the amyloid cascade hypothesis: an update and reappraisal. J Alzheimers Dis 9:151–153
Hardy J, Selkoe DJ (2002) The amyloid hypothesis of Alzheimer's disease: progress and problems on the road to therapeutics. Science 297:353–356
Harris JA, Devidze N, Verret L et al (2010) Transsynaptic progression of amyloid-β-induced neuronal dysfunction within entorhinal-hippocampal network. Neuron 68:428–441
Heimer L, van Hoesen GW (2006) The limbic lobe and its output channels: implications for emotional functions and adaptive behavior. Neurosci Biobehav Rev 30:126–147
Heimer L, de Olmos J, Alheid GF, Záborszky L (1991) "Perestroika" in the basal forebrain: opening the border between neurology and psychiatry. Prog Brain Res 87:109–165
Hertz L (1989) Is Alzheimer's disease an anterograde degeneration, originating in the brainstem, and disrupting metabolic interactions between neurons and glial cells? Brain Res Rev 14:335–353
Higuchi M, Lee MY, Trojanowski JQ (2002) Tau and axonopathy in neurodegenerative disorders. NeuroMol Med 2:131–150
Hof PR, Bussière T, Gold G et al (2003) Stereologic evidence for persistence of viable neurons in layer II of the entorhinal cortex and the CA1 field in Alzheimer disease. J Neuropathol Exp Neurol 62:55–67
Holmes BB, Furman JL, Mahan TE et al (2014) Proteopathic tau seeding predicts tauopathy in vivo. Proc Natl Acad Sci USA 111:E4376–4385
Holstege G, Mouton LJ, Gerrits NM (2004) Emotional motor system. In: Paxinos G, Mai JK (eds) The human nervous system, 2nd edn. Elsevier, San Diego, CA, pp 1306–1325
Howard A, Tamas G, Soltesz I (2005) Lighting the chandelier: new vistas for axo-axonic cells. Trends Neurosci 28:310–316
Hunter JM, Kwan J, Malek-Ahmadi M et al (2012) Morphological and pathological evolution of the brain microcirculation in aging and Alzheimer's disease. PLoS One 7:e36893
Hyman BT, Gómez-Isla T (1994) Alzheimer's disease is a laminar, regional, and neural system specific disease, not a global brain disease. Neurobiol Aging 15:353–354
Hyman BT, Gómez-Isla T (1997) The natural history of Alzheimer neurofibrillary tangles and amyloid deposits. Neurobiol Aging 18:386–387
Hyman BT, Trojanowski JQ (1997) Editorial on consensus recommendations for the postmortem diagnosis of Alzheimer disease from the National Institute on Aging and the Reagan Institute Working Group on diagnostic criteria for the neuropathological assessment of Alzheimer disease. J Neuropathol Exp Neurol 56:1095–1097
Hyman BT, Kromer LJ, van Hoesen GW (1988) A direct demonstration of the perforant pathway terminal zone in Alzheimer's disease using the monoclonal antibody Alz-50. Brain Res 450:392–397
Hyman BT, van Hoesen GW, Damasio AR (1990) Memory-related systems in Alzheimer's disease: an anatomic study. Neurology 40:1721–1730
Hyman BT, Marzloff K, Arrigada PV (1993) The lack of accumulation of senile plaques or amyloid burden in Alzheimer's disease suggests a dynamic balance between amyloid deposition and resolution. J Neuropathol Exp Neurol 52:594–600
Iadecola C (2004) Neurovascular regulation in the normal brain and in Alzheimer's disease. Nat Rev Neurosci 5:347–360

Iadecola C (2010) The overlap between neurodegenerative and vascular factors in the pathogenesis of dementia. Acta Neuropathol 120:287–296

Iba M, Guo JL, McBride JD et al (2013) Synthetic tau fibrils mediate transmission of neurofibrillary tangles in a transgenic mouse model of Alzheimer-like tauopathy. J Neurosci 33: 1024–1037

Ikonomovic MD, Klunk WE, Abrahamson EE et al (2008) Post-mortem correlates of in vivo PiB-PET amyloid imaging in a typical case of Alzheimer's disease. Brain 131:1630–1645

Insausti R, Amaral DG (2012) Hippocampal formation. In: Mai JK, Paxinos G (eds) The human nervous system, 3rd edn. Academic Press, San Diego, CA, pp 896–942

Iqbal K, Grundke-Iqbal I (2008) Alzheimer neurofibrillary degeneration: significance, etiopathogenesis, therapeutics and prevention. J Cell Mol Med 12:38–55

Iqbal K, Grundke-Iqbal I (2011) Opportunities and challenges in developing Alzheimer disease therapeutics. Acta Neuropathol 122:543–549

Iqbal K, Liu F, Gong CX et al (2009) Mechanisms of tau-induced neurodegeneration. Acta Neuropathol 118:53–69

Iseki E, Tsunoda S, Suzuki K et al (2002) Regional quantitative analysis of NFT in brains of non-demented elderly persons: comparisons with findings in brains of late-onset Alzheimer's disease and limbic NFT dementia. Neuropathology 22:34–39

Issidorides MR, Mytilineou C, Panayotacopoulou T, Yahr MD (1991) Lewy bodies in parkinsonism share components with intraneuronal protein bodies of normal brains. J Neural Transm 3:49–61

Ittner A, Ke JD, van Eersel J et al (2011) Brief update on different roles of tau in neurodegeneration. IUBMB Life 63:495–502

Jack CR Jr (2012) Alzheimer disease: new concepts on its neurobiology and the clinical role imaging will play. Radiology 263:344–361

Jack CR Jr, Holtzman DM (2013) Biomarker modelling of Alzheimer's disease. Neuron 80:1347–1358

Jack CR Jr, Knopman DS, Jagust WJ et al (2013) Tracking pathophysiological processes in Alzheimer's disease: an updated hypothetical model of dynamic biomarkers. Lancet Neurol 12:207–216

Jagust WJ, Landau SM, Alzheimer's Disease Neuroimaging Initiative (2012) Apolipoprotein E, not fibrillary β-amyloid, reduces cerebral glucose metabolism in normal aging. J Neurosci 32:18227–18233

Jellinger KA, Attems J (2007) Neurofibrillary tangle-predominant dementia: comparison with classical Alzheimer disease. Acta Neuropathol 113:107–117

Jensen JR, Cisek K, Funk KE et al (2011) Research towards tau imaging. J Alzheimers Dis 26 (Suppl 3):147–157

Johnson SB, Blum RW, Giedd JN (2009) Adolescent maturity and the brain: the promise and pitfalls of neuroscience research in adolescent health policy. J Adolesc Health 45:216–221

Jucker M, Walker LC (2011) Pathogenic protein seeding in Alzheimer's disease and other neurodegenerative disorders. Ann Neurol 70:532–540

Kalaria RN, Stockmeier CA, Harik SI (1989) Brain microvessels are innervated by locus ceruleus noradrenergic neurons. Neurosci Lett 97:203–208

Kalus P, Braak H, Braak BJ (1989) The presubicular region in Alzheimer's disease: topography of amyloid deposits and neurofibrillary changes. Brain Res 494:198–203

Karran E, Mercken M, de Strooper B (2011) The amyloid cascade hypothesis for Alzheimer's disease: an appraisal for the development of therapeutics. Nat Rev Drug Discov 10:698–712

Kaufman SK, Diamond MI (2013) Prion-like propagation of protein aggregation and related therapeutic strategies. Neurotherapeutics 10:371–382

Kemper TL (1984) Neuroanatomical and neuropathological changes in normal aging and in dementia. In: Albert ML (ed) Clinical neurology of aging. Oxford University Press, New York, NY, pp 9–52

Kepe V, Moghbei MC, Långström B et al (2013) Amyloid-β positron emission tomography imaging probes: a critical review. J Alzheimers Dis 36:613–631

Kitt CA, Price DL, Struble RG et al (1985a) Evidence for cholinergic neurites in senile plaques. Science 226:1443–1445

Kitt CA, Struble RG, Cork LC et al (1985b) Catecholaminergic neurites in senile plaques in prefrontal cortex of aged nonhuman primates. Neuroscience 16:691–699

Klunk WE, Engler H, Nordberg A et al (2004) Imaging brain amyloid in Alzheimer's disease with Pittsburgh Compound-B. Ann Neurol 55:306–319

Kolarova M, Garcia-Sierra F, Bartos A et al (2012) Structure and pathology of tau protein in Alzheimer disease. Int J Alzheimer Dis 2012, 731526

Koo EH, Sisodia SS, Archer DR et al (1990) Precursor of amyloid protein in Alzheimer disease undergoes fast axonal transport. Proc Natl Acad Sci USA 87:1561–1565

Kopeikina KJ, Hyman BT, Spires-Jones TL (2012) Soluble forms of tau are toxic in Alzheimer's disease. Transl Neurosci 3:223–233

Köpke E, Tung YC, Shaikh S et al (1993) Microtubule-associated protein tau – abnormal phosphorylation of a non-paired helical filament pool in Alzheimer's disease. J Biol Chem 268:24374–24384

Korczyn AD (2008) The amyloid cascade hypothesis. Alzheimers Dement 4:176–178

Korczyn AD (2013) Is Alzheimer's disease a homogeneous disease entity? J Neural Transm 120:1475–1477

Korczyn AD, Vakhapova V, Grinberg LT (2012) Vascular dementia. J Neurol Sci 322:2–10

Kovacech B, Skrabana R, Novak M (2010) Transition of tau protein from disordered to misordered in Alzheimer's disease. Neurodegener Dis 7:24–27

Kovács T (2013) The olfactory system in Alzheimer's disease: pathology, pathophysiology, and pathway for therapy. Transl Neurosci 4:34–45

Kovács T, Cairns NJ, Lantos PL (1999) β-amyloid deposition and neurofibrillary tangle formation in the olfactory bulb in ageing and Alzheimer's disease. Neuropathol Appl Neurobiol 25:481–491

Kovács T, Milenkovic I, Wöhrer A et al (2013) Non-Alzheimer neurodegenerative pathologies and their combinations are more frequent than commonly believed in the elderly brain: a community-based autopsy series. Acta Neuropathol 126:365–384

Kuchibhotla KV, Wegmann S, Kopeikina KJ et al (2014) Neurofibrillary tangle-bearing neurons are functionally integrated in cortical circuit *in vivo*. Proc Natl Acad Sci USA 111:510–514

Kurylo DD, Corkin S, Allard T et al (1993) Auditory function in Alzheimer's disease. Neurology 43:1893–1899

Lasagna-Reeves C, Castillo-Carranza D, Sengupta U et al (2012) Identification of oligomers at early stages of tau aggregation in Alzheimer's disease. FASEB J 26:1946–1959

Lazarov O, Morfini GA, Lee EB et al (2005) Axonal transport, amyloid precursor protein, kinesin-1, and the processing apparatus: revisited. J Neurosci 25:2386–2395

Lee VM-Y, Goedert M, Trojanowski JQ (2001) Neurodegeneration and tauopathies. Annu Rev Neurol 24:1121–1159

Lee SJ, Desplats P, Sigurdson C et al (2010) Cell-to-cell transmission of non-prion protein aggregates. Nat Rev Neurol 6:702–706

Lee S, Kim W, Li Z, Hall FG (2012a) Accumulation of vesicle-associated human tau in distal dendrites drives degeneration and tau secretion in an in situ cellular tauopathy model. Int J Alzheimers Dis 2012, 172837

Lee Y, Morrison BM, Li Y, Lengacher S et al (2012b) Oligodendroglia metabolically support axons and contribute to neurodegeneration. Nature 487:443–448

Li B, Chohan MO, Grundke-Iqbal I, Iqbal K (2007) Disruption of microtubule network by Alzheimer abnormally hyperphosphorylated tau. Acta Neuropathol 113:501–511

Liberati G, Raffone A, Belardinelli O (2012) Cognitive reserve and its implications for rehabilitation and Alzheimer's disease. Cogn Process 13:1–12

Linn RT, Wolf PA, Bachman DL et al (1995) The 'preclinical phase' of probably Alzheimer's disease. A 13-year prospective study of the Framingham cohort. Arch Neurol 52:485–490
Liu L, Drouet V, Wu JW et al (2012) Trans-synaptic spread of tau pathology in vivo. PLoS One 7: e31302
Lou K, Yao Y, Hoye AT, James MJ et al (2014) Brain-penetrant, orally bioavailable microtubule-stabilizing small molecules are potential candidate therapeutics for Alzheimer's disease and related tauopathies. J Med Chem 57:6116–6127
Love S, Chalmers K, Ince P et al (2014) Development, appraisal, validation and implementation of a consensus protocol for the assessment of cerebral amyloid angiopathy in post-mortem brain tissue. Am J Neurodegener Dis 3:19–32
Maarouf CL, Daugs ID, Kokjohn TA et al (2010) The biochemical aftermath of anti-amyloid immunotherapy. Mol Neurodegener 5:39–54
Mackic JB, Bading J, Ghiso J et al (2002) Circulating amyloid-beta peptide crosses the blood-brain barrier in aged monkeys and contributes to Alzheimer's disease lesions. Vascul Pharmacol 38:303–313
Maeda S, Sahara N, Saito Y et al (2007) Granular tau oligomers as intermediates of tau filaments. Biochemistry 46:3856–3861
Mandelkow EM, Mandelkow E (1998) Tau in Alzheimer's disease. Trends Cell Biol 8:425–427
Mandelkow EM, Mandelkow E (2012) Biochemistry and cell biology of tau protein in neurofibrillary degeneration. Cold Spring Harb Perspect Med 2:a006247
Mandelkow E, von Bergen M, Biernat J, Mandelkow EM (2007) Structural principles of tau and the paired helical filaments of Alzheimer's disease. Brain Pathol 17:83–90
Mann DM (1983) The locus coeruleus and its possible role in ageing and degenerative disease of the human central nervous system. Mech Ageing Dev 23:73–94
Mann DMA, Hardy J (2013) Amyloid or tau – the chicken or the egg? Acta Neuropathol 126:609–613
Mann DMA, Yates PO, Hawkes J (1982) The noradrenergic system in Alzheimer and multi-infarct dementias. J Neurol Neurosurg Psychiatry 45:113–119
Marien MR, Colpaert FC, Rosenquist AC (2004) Noradrenergic mechanisms in neurodegenerative diseases: a theory. Brain Res Brains Res Rev 45:38–78
Markesbery WR, Schmitt FA, Kryscio RJ et al (2006) Neuropathologic substrate of mild cognitive impairment. Arch Neurol 63:38–46
Maruyama M, Shimada H, Suhara T et al (2013) Imaging of tau pathology in a tauopathy mouse model and in Alzheimer patients compared to normal controls. Neuron 79:1–15
Masliah E, Alford M, Adame A et al (2003) Abeta1-42 promotes cholinergic sprouting in patients with AD and Lewy body variant of AD. Neurology 61:206–211
Masters CL, Beyreuther K (2006) Pathways to the discovery of the neuronal origin and proteolytic biogenesis of Aβ amyloid of Alzheimer's disease. In: Jucker M, Beyreuther K, Haass C, Nitsch R, Christen Y (eds) Alzheimer: 100 years and beyond. Springer, Berlin, pp 143–149
Masters CL, Selkoe DJ (2012) Biochemistry of amyloid β-protein and amyloid deposits in Alzheimer's disease. Cold Spring Harb Perspect Med 2:a006262
Mattson MP (2004) Pathways towards and away from Alzheimer's disease. Nature 430:631–639
Mattsson N, Zetterberg H, Hansson O et al (2009) CSF biomarkers and incipient Alzheimer disease in patients with mild cognitive impairment. JAMA 302:385–393
Mattsson N, Portelius E, Rolstad S et al (2012) Longitudinal cerebrospinal fluid biomarkers over four years in mild cognitive impairment. J Alzheimers Dis 30:767–778
Mayeux R (2010) Early Alzheimer's disease. N Engl J Med 362:2194–2201
Mayeux R, Stern Y (2012) Epidemiology of Alzheimer's disease. Cold Spring Harb Perspect Med 2:a006239
Medina M, Ávila J (2014a) The role of extracellular tau in the spreading of neurofibrillary pathology. Front Cell Neurosci 8:113
Medina M, Ávila J (2014b) New perspectives on the role of tau in Alzheimer's disease. Implication for therapy. Biochem Pharmacol 88:540–547

Meel-van den Abeelen AS, Lagro J et al (2013) Baroreflex function is reduced in Alzheimer's disease: a candidate biomarker? Neurobiol Aging 34:1170–1176

Meng X, D'Arcy C (2012) Education and dementia in the context of the cognitive reserve hypothesis: a systematic review with meta-analyses and quantitative analyses. PLoS One 7: e38268

Mercken M, Vandermeeren M, Lübke U et al (1992) Monoclonal antibodies with selective specifity for Alzheimer tau are directed against phosphatase-sensitive epitopes. Acta Neuropathol 84:265–272

Merino-Serrais P, Benavides-Piccione R, Blazquez-Llorca L et al (2013) The influence of phospho-tau on dendritic spines of cortical pyramidal neurons in patients with Alzheimer's disease. Brain 136:1913–1928

Mesulam MM (1998) From sensation to cognition. Brain 121:1013–1052

Mesulam MM (2004) The cholinergic innervation of the human cerebral cortex. Prog Brain Res 145:67–78

Mesulam MM, Mufson EJ (1993) The insula of Reil in man and monkey. In: Jones EG, Peters A (eds) Cerebral cortex. Association and auditory cortices, vol 4. Plenum, New York, NY, pp 179–225

Mesulam MM, Shaw P, Mash D, Weintraub S (2004) Cholinergic nucleus basalis tauopathy emerges early in the aging-MCI-AD continuum. Ann Neurol 55:815–828

Mesulam MM, Weintraub S, Rogalski EJ et al (2014) Asymmetry and heterogeneity of Alzheimer's and frontotemporal pathology in primary progressive aphasia. Brain 137:1176–1192

Mirra S, Heyman A, McKeel D et al (1991) The Consortium to Establish a Registry for Alzheimer's Disease (CERAD). Part II. Standardization of the neuropathologic assessment of Alzheimer's disease. Neurology 41:479–486

Moceri VM, Kukull WA, Emanuel I et al (2000) Early-life risk factors and the development of Alzheimer's disease. Neurology 54:415–420

Molnar FJ, Man-Son-Hing M, Fergusson D (2009) Systematic review of measures of clinical significance employed in randomized controlled trials of drugs for dementia. J Am Geriatr Soc 57:536–546

Montine TJ, Phelps CH, Beach TG et al (2012) National Institute on Aging-Alzheimer's Association guidelines for the neuropathologic assessment of Alzheimer's disease: a practical approach. Acta Neuropathol 123:1–11

Morawski M, Brückner G, Jäger C et al (2010) Neurons associated with aggrecan-based perineuronal nets are protected against tau pathology in subcortical regions in Alzheimer's disease. Neuroscience 169:1347–1363

Morris JC, Storandt M, Miller JP et al (2001) Mild cognitive impairment represents early-stage Alzheimer disease. Arch Neurol 58:397–405

Morris JC, Blennow K, Froelich L et al (2014) Harmonized diagnostic criteria for Alzheimer's disease: recommendations. J Intern Med 275:204–210

Morrison JH, Hof PR (1997) Life and death of neurons in the aging brain. Science 278:412–419

Morrison JH, Foote SL, O'Connor D, Bloom FE (1982) Laminar, tangential and regional organization of the nordrenergic innervation of monkey cortex: dopamin-β-hydroxylase immunohistochemistry. Brain Res Bull 9:309–319

Morsch R, Simon W, Coleman PD (1999) Neurons may live for decades with neurofibrillary tangles. J Neuropathol Exp Neurol 58:188–197

Mount C, Downton C (2006) Alzheimer disease: progress or profit? Nat Med 12:780–784

Mucke L, Selkoe DJ (2012) Neurotoxicity of amyloid β-protein: synaptic and network dysfunction. Cold Spring Harb Perspect Med 2:a006338

Mufson EJ, Chen EY, Cochran EJ et al (1999) Entorhinal cortex β-amyloid load in individuals with mild cognitive imüairment. Exp Neurol 158:469–490

Mufson EJ, Binder L, Counts SE et al (2012) Mild cognitive impairment: pathology and mechanisms. Acta Neuropathol 123:13–30

Mufson EJ, Ward S, Binder L (2013) Prefibrillar tau oligomers in mild cognitive impairment and Alzheimer's disease. Neurodegener Dis 13:151–153
Munoz DG, Wang D (1992) Tangle-associated neuritic clusters. A new lesion in Alzheimer's disease and aging suggests that aggregates of dystrophic neurites are not necessarily associated with beta/A4. Am J Pathol 140:1167–1178
Muresan Z, Muresan V (2008) Seeding neuritic plaques from the distance: a possible role for brainstem neurons in the development of Alzheimer's disease pathology. Neurodegener Dis 5:250–253
Nave KA (2010) Myelination and support of axonal integrity by glia. Nature 468:244–252
Nelson PT, Jicha GA, Schmitt FA et al (2007) Clinicopathologic correlations in a large Alzheimer disease center autopsy cohort: neuritic plaques and neurofibrillary tangles "do count" when staging disease severity. J Neuropathol Exp Neurol 66:1136–1146
Nelson PT, Braak H, Markesbery WR (2009) Neuropathology and cognitive impairment in Alzheimer's disease: a complex but coherent relationship. J Neuropathol Exp Neurol 68:1–14
Nelson PT, Head E, Schmitt FA et al (2011) Alzheimer's disease is not "brain aging": neuropathological, genetic, and epidemiologicical human studies. Acta Neuropathol 121:571–587
Niblock M, Gallo JM (2012) Tau alternative splicing in familial and sporadic tauopathies. Biochem Soc Trans 40:677–680
Nieuwenhuys R (1994) The neocortex. An overview of its evolutionary development, structural organization and synaptology. Anat Embryol 190:307–337
Nieuwenhuys R (1996) The greater limbic system, the emotional motor system and the brain. Prog Brain Res 107:551–580
Nieuwenhuys R (1999) Structure and organisation of fibre systems. In: Nieuwenhuys R, Ten Donkelaar JH, Nicholson C (eds) The central nervous system of vertebrates, vol 1. Springer, Berlin, pp 113–157
Nieuwenhuys R (2012) The insular cortex: a review. Prog Brain Res 195:123–163
O'Donnell J, Zeppenfeld D, McConnell E et al (2012) Norepinephrine: a neuromodulator that boosts the function of multiple cell types to optimize CNS performance. Neurochem Res 37:2496–2512
Ohm TG, Müller H, Braak H, Bohl J (1995) Close-meshed prevalence rates of different stages as a tool to uncover the rate of Alzheimer's disease-related neurofibrillary changes. Neuroscience 64:209–217
Okamura N, Furumoto S, Harada R et al (2013) Novel 18F-labeled arylquinoline derivatives for noninvasive imaging of tau pathology in Alzheimer disease. J Nucl Med 54:1420–1427
Pamphlett R (2014) Uptake of environmental toxicants by the locus ceruleus. A potential trigger for enruodegenerative, demyelinating and psychiatric disorders. Med Hypotheses 82:97–104
Panula P, Airaksinen MS, Pirvola U, Kotilainen E (1990) A histamine-containing neuronal system in human brain. Neuroscience 34:127–132
Parvizi J, van Hoesen GW, Damasio A (1998) Severe pathological changes of parabrachial nucleus in Alzheimer's disease. Neuroreport 9:4151–4154
Parvizi J, van Hoesen GW, Damasio A (2001) The selective vulnerability of brainstem nuclei to Alzheimer's disease. Ann Neurol 49:53–66
Pearson RCA (1996) Cortical connections and the pathology of Alzheimer's disease. Neurodegeneration 5:429–434
Pearson RCA, Powell TPS (1989) The neuroanatomy of Alzheimer's disease. Rev Neurosci 2:101–122
Perani D (2014) FDG-PET and amyloid-PET imaging: the diverging paths. Curr Opin Neurol 27:405–413
Petersen RC (2000) Mild cognitive impairment: transition between aging and Alzheimer's disease. Neurologia 15:93–101
Petersen RC (2009) Early diagnosis of Alzheimer's disease: is MCI too late? Curr Alzheimer Res 6:324–330
Petersen RC (2011) Clinical practice. Mild cognitive impairment. N Engl J Med 364:2227–2234

Petersen RC, Doody R, Kurz A et al (2001) Current concepts in mild cognitive impairment. Arch Neurol 58:1985–1992

Petrides M, Pandya DN (2004) The frontal cortex. In: Paxinos G, Mai JK (eds) The human nervous system, 2nd edn. Elsevier, San Diego, CA, pp 951–974

Pikkarainen M, Martikainen P, Alafuzoff I (2010) The effect of prolonged fixation time on immunohistochemical staining of common neurodegenerative disease markers. J Neuropathol Appl Neurol 69:40–52

Pimplikar SW (2009) Reassessing the amyloid cascade hypothesis of Alzheimer's disease. Int J Biochem Cell Biol 41:1261–1268

Pohanka M (2013) Alzheimer's disease and oxidative stress: a review. Curr Med Chem 21:356–364

Price JL, Morris JC (1999) Tangles and plaques in nondemented aging and "preclinical" Alzheimer's disease. Ann Neurol 45:358–368

Price JL, Morris JC (2004) So what if tangles precede plaques? Neurobiol Aging 25:721–723

Price JL, Davis PB, Morris JC, White DL (1991) The distribution of tangles, plaques and related immunohistochemical markers in healthy aging and Alzheimer's disease. Neurobiol Aging 12:295–312

Price JL, Carmichael ST, Drevets WC (1996) Networks related to the orbital and medial prefrontal cortex; a substrate for emotional behavior? Prog Brain Res 107:523–526

Prusiner S (2012) Cell biology. A unifying role for prions in neurodegenerative diseases. Science 336:1511–1515

Purohit DP, Batheja NO, Sano M et al (2011) Profiles of Alzheimer's disease-related pathology in an aging urban population sample. J Alzheimer Dis 24:187–196

Purpura DP, Baker HJ (1978) Meganeurites and other aberrant processes of neurons in feline GM1-gangliosidosis: a Golgi study. Brain Res 143:13–26

Qiu C, Kivipelto M, von Strauss E (2009) Epidemiology of Alzheimer's disease: occurrence, determinants, and strategies toward intervention. Dialogues Clin Neurosci 11:111–128

Raichle ME, Hartman BK, Eichling JO, Sharpe LG (1975) Central noradrenergic regulation of cerebral blood flow and vascular permeability. Proc Natl Acad Sci USA 72:3726–3730

Rajendran L, Annaert W (2012) Membrane trafficking pathways in Alzheimer's disease. Traffic 13:759–770

Ramirez MJ, Lai MK, Tordera RM, Francis PT (2014) Serotonergic therapies for cognitive symptoms in Alzheimer's disease: rationale and current status. Drugs 74:729–736

Ramón-Moliner E, Nauta WJH (1966) The isodendritic core of the brain stem. J Comp Neurol 126:311–336

Rapoport SI (1988) Brain evolution and Alzheimer's disease. Rev Neurol (Paris) 144:79–90

Rapoport SI (1989) Hypothesis: Alzheimer's disease is a phylogenetic disease. Med Hypotheses 29:147–150

Rapoport SI (1990) Integrated phylogeny of the primate brain, with special reference to humans and their diseases. Brain Res Rev 15:267–294

Rapoport SI (1999) How did the human brain evolve? A proposal based on new evidence from *in vivo* imaging during attention and ideation. Brain Res Bull 50:149–165

Rapoport SI, Nelson PT (2011) Biomarkers and evolution in Alzheimer's disease. Prog Neurobiol 95:510–513

Reid AT, Evans AC (2013) Structural networks in Alzheimer's disease. Eur Neuropsychopharmacol 23:63–77

Reinhard JF, Liebmann JE, Schlosberg AL, Moskowitz MA (1979) Serotonin neurons project to small blood vessels in the brain. Science 206:86–87

Reisberg B, Franssen EH, Hasan SM et al (1999) Retrogenesis: clinical, physiologic, and pathologic mechanisms in brain aging, Alzheimer's and other dementing processes. Eur Arch Psychiatry Clin Neurosci 249(Suppl 3):28–36

Reisberg B, Franssen EH, Souren LE et al (2002) Evidence and mechanisms of retrogenesis in Alzheimer's and other dementias: management and treatment import. Am J Alzheimers Dis Other Demen 17:202–212

Reitz C, Brayne C, Mayeux R (2011) Epidemiology of Alzheimer disease. Nat Rev Neurol 7:137–152

Rentz DM, Parra Rodriguez MA et al (2013) Promising developments in neuropsychological approaches for the detection of preclinical Alzheimer's disease: a selective review. Alzheimers Res Ther 5:58

Revesz T, Holton JL, Lashley T et al (2009) Genetics and molecular pathogenesis of sporadic and hereditary cerebral amyloid angiopathies. Acta Neuropathol 118:115–130

Rockland KS, Pandya DN (1979) Laminar origins and terminations of cortical connections of the occipital lobe in the rhesus monkey. Brain Res 179:3–20

Rosa MI, Perucchi J, Medeiros LR et al (2014) Accuracy of cerebrospinal fluid Aβ(1-42) for Alzheimer's disease diagnosis: a systematic review and meta-analysis. J Alzheimers Dis 40:443–454

Rosén C, Zetterberg H (2013) Cerebrospinal fluid biomarkers for pathological processes in Alzheimer's disease. Curr Opin Psychiatry 26:276–282

Royall DR (2008) Insular Alzheimer disease pathology and the psychometric correlates of mortality. Cleve Clin J Med 75(Suppl 2):97–99

Royall DR, Gao JH, Kellogg DL (2008) Insular Alzheimer's disease pathology as a cause of 'age-related' autonomic dysfunction and mortality in the non-demented elderly. Med Hypotheses 67:747–758

Rüb U, Del Tredici K, Schultz C et al (2000) The evolution of Alzheimer's disease-related cytoskeletal pathology in the human raphe nuclei. Neuropathol Appl Neurobiol 26:553–557

Rüb U, Del Tredici K, Schultz C et al (2001a) The premotor region essential for rapid vertical eye movements shows early involvement in Alzheimer's disease-related cytoskeletal pathology. Vision Res 41:2149–2156

Rüb U, Del Tredici K, Schultz C et al (2001b) The autonomic higher oder processing nuclei of the lower brain stem are among the early targets of the Alzheimer's disease-related cytoskeletal pathology. Acta Neuropathol 101:555–564

Rubial-Álvarez S, de Sola S, Machado MC et al (2013) The comparison of cognitive and functional performance in children and Alzheimer's disease supports the retrogenesis model. J Alzheimers Dis 33:193–203

Run X, Liang Z, Zhang L et al (2009) Anesthesia induces phosphorylation of tau. J Alzheimer Dis 16:619–629

Sabbagh MN, Cooper K, DeLange J et al (2010) Functional, global and cognitive decline correlates to accumulation of Alzheimer's pathology in MCI and AD. Curr Alzheimer Res 7:280–286

Salloway S, Sperling R, Fox NC et al (2014) Two phase 3 trials of bapineuzumab in mild-to-moderate Alzheimer's disease. N Engl J Med 370:322–333

Sämgård K, Zetterberg H, Blennow K et al (2010) Cerebrospinal fluid total tau as a marker of Alzheimer's disease intensity. Int J Geriatr Psychiatry 25:403–410

Samuels ER, Szabadi E (2008) Functional neuroanatomy of the noradrenergic locus coeruleus: its roles in the regulation of arousal and autonomic function part I: principles of functional organization. Curr Neuropharmacol 6:235–253

Sanders DW, Kaufmann SW, DeVos SL et al (2014) Distinct tau prion strains propagate in cells and mice and define different tauopathies. Neuron 82:1271–1288

Saper CB (2004) Hypothalamus. In: Paxinos G, Mai JK (eds) The human nervous system, 2nd edn. Elsevier, San Diego, CA, pp 514–550

Saper CB, Weiner BH, German DC (1987) Axonal and transneuronal transport in the transmission of neurological disease: potential role in system degenerations, including Alzheimer's disease. Neuroscience 23:389–398

Sara SJ (2009) The locus coeruleus and noradrenergic modulation of cognition. Nat Rev Neurosci 10:211–223

Sassin I, Schultz C, Thal DR et al (2000) Evolution of Alzheimer's disease-related cytoskeletal changes in the basal nucleus of Meynert. Acta Neuropathol 100:259–269

Savica R, Petersen RC (2011) Prevention of dementia. Psychiatr Clin North Am 34:127–145

Schlaug G, Norton A, Overy K et al (2003) Effects of music training on the child's brain and cognitive development. Ann NY Acad Sci 1060:219–230

Schliebs R, Arendt T (2011) The cholinergic system in aging and neuronal degeneration. Behav Brain Res 221:555–563

Scholz T, Mandelkow E (2014) Transport and diffusion of tau in neurons. Cell Mol Life Sci 71:3139–3150

Schönheit B, Zarski R, Ohm TG (2004) Spatial and temporal relationship between plaques and tangles in Alzheimer-pathology. Neurobiol Aging 25:697–711

Seitz DP, Reimer CL, Siddiqui N (2012) A review of epidemiological evidence for general anaesthesia as a risk factor for Alzheimer's disease. Prog Neuropsychoüharmacol Biol Psychiatry 47:122–127

Selkoe DJ (1994) Alzheimer's disease: a central role for amyloid. J Neuropathol Exp Neurol 53:438–447

Selkoe DJ (2000) Toward a comprehensive theory for Alzheimer's disease: hypothesis: Alzheimer's disease is caused by the cerebral accumulation and cytotoxicity of amyloid beta-protein. Ann NY Acad Sci 924:17–25

Selkoe D, Mandelkow E, Holtzman D (2012) Deciphering Alzheimer disease. Cold Spring Harb Perspect Med 2:1–8

Serrano-Pozo A, Frosch M, Masliah E, Hyman BT (2011) Neuropathological alterations in Alzheimer disease. Cold Spring Harb Perspect Med 1:a006189

Serrano-Pozo A, Quian J, Monsell SE et al (2013) Examination of the clinicopathological continuum of Alzheimer disease in the autopsy cohort of the National Alzheimer Coordinating Center. J Neuropathol Exp Neurol 72:1182–1192

Shaw P, Kabani HJ, Lerch JP et al (2008) Neurodevelopmental trajectories of the human cerebral cortex. J Neurosci 28:3586–3594

Silverman W, Wisniewski HM, Bobinski M, Wegiel J (1997) Frequency of stages of Alzheimer-related lesions in different age categories. Neurobiol Aging 18:377–379

Simic G, Bexheti S, Kelovic Z et al (2005) Hemispheric asymmetry, modular variability and age-related changes in the human entorhinal cortex. Neuroscience 130:911–925

Simic G, Stanic G, Mladinov M et al (2009) Does Alzheimer's disease begin in the brainstem? Neuropathol Appl Neurobiol 35:532–554

Sinha UK, Hollen KM, Rodriguez R, Miller CA (1993) Auditory system degeneration in Alzheimer's disease. Neurology 43:779–785

Smith GS, Kramer E, Ma Y et al (2009) Cholinergic modulation of the cerebral metabolic response to citaopram in Alzheimer's disease. Brain 132:392–401

Sperling RA, Aisen PS, Beckett LA et al (2011) Toward defining the preclinical stages of Alzheimer's disease: recommendations from the National Institute on Aging-Alzheimer's Association workgroups on diagnostic guidelines for Alzheimer's disease. Alzheimers Dement 7:280–292

Spillantini MG, Goedert M (2013) Tau pathology and neurodegeneration. Lancet Neurol 12:609–622

Squire LR, Schacter DL (2002) Neuropsychology of memory. The Guilford, New York, NY

Stephan H (1983) Evolutionary trends in limbic structures. Neurosci Biobehav Rev 7:367–374

Stern Y (2002) What is cognitive reserve? Theory and research application of the reserve concept. J Int Neuropsychol Soc 8:448–460

Stern Y (2009) Cognitive reserve. Neuropsychologia 47:2015–2028

Stokin GB, Goldstein LSB (2006) Axonal transport and Alzheimer's disease. Ann Rev Biochem 75:607–627

Streit WJ, Braak H, Xue QS, Bechmann I (2009) Dystrophic (senescent) rather than activated microglial cells are associated with tau pathology and likely precede neurodegeneration in Alzheimer's disease. Acta Neuropathol 118:475–485

Streit WJ, Xue QS, Braak H, Del Tredici K (2014) Presence of severe neuroinflammation does not intensify neurofibrillary degeneration in human brain. Glia 62:96–105

Stricker NH, Schweinsburg BC, Delano-Wood L et al (2009) Decreased white matter integrity in late-myelinating fiber pathways in Alzheimer's disease supports retrogenesis. Neuroimage 45:10–16

Struble RG, Price DL Jr, Cork LC, Price DL (1985) Senile plaques in cortex of aged normal monkeys. Brain Res 361:267–275

Suzuki WA, Amaral DG (2004) Functional neuroanatomy of the medial temporal lobe memory system. Cortex 40:220–222

Szabadi E (2013) Functional neuroanatomy of the central noradrenergic system. J Psychopharmacol 27:659–693

Tago T, Furumoto S, Okamura N et al (2014) Synthesis and preliminary evaluation of 2-arylhydroxyquinoline derivatives for tau imaging. J Label Compd Radiopharm 57:18–24

Tapiola T, Alafuzoff I, Herukka SK et al (2009) Cerebrospinal fluid (beta)-amyloid 42 and tau proteins as biomarker changes in the brain. Arch Neurol 66:382–389

Tayab HO, Yang HD, Price BH, Tarazi FI (2012) Pharmacotherapies for Alzheimer's disease: beyond cholinesterase inhibitors. Pharmacol Ther 134:8–25

Taylor KI, Probst A (2008) Anatomic localization of the transentorhinal region of the perirhinal cortex. Neurobiol Aging 29:1591–1596

Thal DR, Sassin I, Schultz C et al (1999) Fleecy amyloid deposits in the internal layers of the human entorhinal cortex are comprised of N-terminal truncated fragments of Abeta. J Neuropathol Exp Neurol 58:210–216

Thal DR, Rüb U, Orantes M, Braak H (2002) Phases of Aβ-deposition in the human brain and its relevance for the development of AD. Neurology 58:1791–1800

Thal DR, Ghebremedhin E, Orantes M, Wiestler OD (2003) Vascular pathology in Alzheimer's disease: correlation cerebral amyloid angiopathy and arteriosclerosis/lipohyalinosis with cognitive decline. J Neuropathol Exp Neurol 62:1287–1301

Thal DR, Grinberg LT, Attems J (2012) Vascular dementia: different forms of vessel disorders contribute to the development of dementia in the elderly brain. Exp Gerontol 47:816–824

Thal DR, von Arnim C, Griffin WS et al (2013) Pathology of clinical and preclinical Alzheimer's disease. Eur Arch Psychiatry Clin Neurosci 263(Suppl 2):137–145

Thies E, Mandelkow EM (2007) Missorting of tau in neurons causes degeneration of synapses that can be rescued by the kinase MARK2/Par-1. J Neurosci 27:2896–2907

Tian H, Davidowitz E, Lopez P et al (2013) Trimeric tau is toxic to human neuronal cells at low nanomolar concentrations. Int J Cell Biol 2013, 260787

Tiraboschi P, Hansen LA, Thal LJ, Corey-Bloom J (2004) The importance of neuritic plaques and tangles to the development and evolution of AD. Neurology 62:1984–1989

Togo T, Akiyama H, Iseki E et al (2004) Immunohistochemical study of tau accumulation in early stages of Alzheimer-type neurofibrillary lesions. Acta Neuropathol 107:504–508

Toledo JB, Brettschneider J, Grossman M et al (2012) CSF biomarkers cutoffs: the importance of coincidental neuropathological diseases. Acta Neuropathol 124:23–35

Tolnay M, Probst A (1999) Review: tau protein pathology in Alzheimer's disease and related disorders. Neuropathol Appl Neurobiol 25:171–187

Tomlinson BE, Irving D, Blessed G (1981) Cell loss in the locus coeruleus in senile dementia of Alzheimer type. J Neurol Sci 49:419–428

Toussay X, Basu K, Lacoste B, Hamel E (2013) Locus coeruleus stimulation recruits a broad cortical neuronal network and increases cortical perfusion. J Neurosci 33:3390–3401

Trillo L, Das D, Hsieh W et al (2013) Ascending monoaminergic systems alterations in Alzheimer's disease, translating basic science into clinical care. Neurosci Biobehav Rev 37:1363–1379

Trojanowski JQ, Hampel H (2011) Neurogenerative disease biomarkers: guideposts for disease prevention through early diagnosis and intervention. Prog Neurobiol 95:491–495
Trojanowski JQ, Lee VMY (2000) "Fatal Attractions" of proteins: a comprehensive hypothetical mechanism underlying Alzheimer's disease and other neurodegenerative disorders. Ann NY Acad Sci 924:62–67
Trojanowski JQ, Schuck T, Schmidt ML, Lee VM (1989) Distribution of tau proteins in the normal human central and peripheral nervous system. J Histochem Cytochem 37:209–215
Uchihara T, Nakamura A, Yamazaki M, Mori O (2001) Evolution from pretangle neurons to neurofibrillary tangles monitored by thiazin red combined with Gallyas method and double immunofluorescence. Acta Neuropathol 101:535–539
Uchihara T, Shibuya K, Nakamura A, Yagishita S (2005) Silver stains distinguish tau-positive structures in corticobasal degeneration/progressive supranuclear palsy and in Alzheimer's disease – comparison between Gallyas and Campbell-Switzer methods. Acta Neuropathol 109:299–305
Uchihara T, Nakamura A, Shibuya K, Yagishita S (2011) Specific detection of pathological three-repeat tau after pretreatment with potassium permanganate and oxalic acid in PSP/CBD brains. Brain Pathol 21:180–188
Uchihara T, Hara M, Nakamura A, Hirokawa K (2012) Tangle evolution linked to differential 3- and 4-repeat tau isoform deposition: a double immunofluorolabeling study using two monoclonal antibodies. Histochem Cell Biol 137:261–267
Ullén F (2009) Is activity regulation of late myelination a plastic mechanism in the human nervous system? Neuron Glia Biol 5:29–34
Usher M, Cohen JD, Servan-Schreiber D et al (1999) The role of locus coeruleus in the regulation of cognitive performance. Science 283:549–554
Van Ba AT, Imberdis T, Perrier V (2013) From prion disease to prion-like propagation mechanisms of neurodegenerative diseases. Int J Cell Biol 2013, 975832
van der Knaap MS, Valk J, Bakker CJ et al (1991) Myelination as an expression of the functional maturity of the brain. Dev Med Child Neurol 33:849–857
van der Werf WMP, Groenewegen HJ (2002) The intralaminar and midline nuclei of the thalamus. Anatomical and functional evidence for participation in processes of arousal and awareness. Brain Res Rev 39:107–140
van Gool WA, Eikelenboom P (2000) The two faces of Alzheimer's disease. J Neurol 247:500–505
van Hoesen GW, Hyman BT (1990) Hippocampal formation: anatomy and the patterns of pathology in Alzheimer's disease. Prog Brain Res 83:445–457
Vana L, Kanaan NM, Ugwu IC et al (2011) Progression of tau pathology in cholinergic basal forebrain neurons in mild cognitive impairment and Alzheimer's disease. Am J Pathol 179:2533–2550
Vanderstichele HM, Shaw L, Vandijck M et al (2013) Alzheimer disease biomarker testing in cerebrospinal fluid: a method to harmonize assay platforms in the absence of an absolute reference standard. Clin Chem 59:710–712
Velasco ME, Smith MA, Siedlak SI et al (1998) Striation is the characteristic neuritic abnormality in Alzheimer disease. Brain Res 813:329–333
Villemagne VL, Furumoto S, Fodero-Tavoletti MT et al (2014) In vivo evaluation of a novel tau imaging tracer for Alzheimer's disease. Eur J Nucl Med Mol Imaging 41:816–826
Viswanathan A, Greenberg SM (2011) Cerebral amyloid angiopathy in the elderly. Ann Neurol 70:871–880
Vogt BA (2009) Regions and subregions of the cingulate cortex. In: Vogt BA (ed) Cingulate neurobiology and disease. Oxford University Press, New York, NY, pp 3–30
Vogt C, Vogt O (1919) Allgemeinere Ergebnisse unserer Hirnforschung. J Psychol Neurol 25:277–462

Vogt BA, Sikes RW, Vogt LJ (1993) Anterior cingulate cortex and the medial pain system. In: Vogt BA, Gabriel M (eds) Neurobiology of cingulate cortex and limbic thalamus. Birkhäuser, Boston, MA, pp 313–344

von Bergen M, Barghorn S, Biernat J et al (2005) Tau aggregation is driven by a transition from random coil to beta sheet structure. Biochim Biophys Acta 1739:158–166

Vonsattel JP, Myers RH, Hedley-Whyte ET et al (1991) Cerebral amyloid angiopathy without and with cerebral hemorrhages: a comparative histological study. Ann Neurol 30:637–649

Wagner J, Ryazanov S, Leonov A et al (2013) Anle138b: a novel oligomer modulator for disease-modifying therapy of neurodegenerative diseases such as prion and Parkinson's disease. Acta Neuropathol 125:795–813

Walker LC, Diamond MI, Duff KE, Hyman BT (2013) Mechanisms of protein seeding in neurodegenerative diseases. JAMA Neurol 70:304–310

Wang BW, Lu E, Mackenzie IR et al (2012) Multiple pathologies are common in Alzheimer patients in clinical trials. Can J Neurol Sci 39:592–599

Weaver CL, Espinoza M, Kress Y, Davies P (2000) Conformational change as one of the earliest alterations of tau in Alzheimer's disease. Neurobiol Aging 21:719–727

Weinshenker D (2008) Functional consequences of locus coeruleus degeneration in Alzheimer's disease. Curr Alzheimer Res 5:342–345

Weller RO, Massey A, Newman TA et al (1998) Cerebral amyloid angiopathy: amyloid beta accumulates in putative interstitial fluid drainage pathways in Alzheimer's disease. Am J Pathol 153:725–733

Weller RO, Biche D, Nicoll JA (2009) Microvasculature changes and cerebral amyloid angiopathy in Alzheimer's disease and their potential impact on therapy. Acta Neuropathol 118:87–102

Whitehouse PJ, Price DL, Clark AW et al (1981) Alzheimer disease: evidence for selective loss of cholinergic neurons in the nucleus basalis. Ann Neurol 10:122–126

Whitehouse PJ, Struble RG, Hedreen JC et al (1985) Alzheimer's disease and related dementias: selective involvement of specific neuronal systems. CRC Crit Rev Clin Neurobiol 1:319–339

Whittington RA, Bretteville A, Dickler MF, Planel E (2013) Anesthesia and tau pathology. Prog Neuro-Psychopharmacol Biol Psychiatry 47:147–155

Witter MP (1993) Organization of the hippocampal-entorhinal system: a review of current anatomical data. Hippocampus 3 Spec No:33–44

Woodruff AR, Anderson SA, Yuste R (2010) The enigmatic function of chandelier cells. Front Neurosci 4:201

Xu W, Yu JT, Tan MS, Tan L (2014) Cognitive reserve and Alzheimer's disease. Mol Neurobiol. May 4 [Epub ahead of print]

Yamada M, Naiki H (2012) Cerebral amyloid angiopathy. Prog Mol Biol Transl Sci 107:41–78

Yan MH, Wang X, Zhu X (2013) Mitochondrial defects and oxidative stress in Alzheimer disease and Parkinson disease. Free Radic Biol Med 62:90–101

Yilmazer-Hanke DM (2012) Amygdala. In: Mai JK, Paxinos G (eds) The human nervous system, 3rd edn. Academic Press, San Diego, CA, pp 759–835

Yoshida M (2006) Cellular tau pathology and immunohistochemical study of tau isoforms in sporadic tauopathies. Neuropathology 26:457–470

Zaborszky I, Hoemke I, Mohlberg H et al (2008) Stereotaxic probabilistic maps of the magnoscellular cell groups in human basal forebrain. Neuroimage 42:1127–1141

Zakrzewska-Pniewska B, Gawel M, Szmidt-Salkowska E et al (2012) Clinical and functional assessment of dysautonomia and its correlation in Alzheimer's disease. Am J Alzheimers Dis Other Demen 27:592–599

Zanelli O, Solerte SB, Cantoni F (2009) Life expectancy in Alzheimer's disease (AD). Arch Gerontol Geriatr 49(Suppl 1):237–243

Zarow C, Lyness SA, Mortimer JA, Chui HC (2003) Neuronal loss is greater in the locus coeruleus than nucleus basalis and substantia nigra in Alzheimer and Parkinson diseases. Arch Neurol 60:337–341

Zetterberg H, Pedersen M, Lind K et al (2007) Intra-individual stability of CSF biomarkers for Alzheimer's disease over two years. J Alzheimers Dis 12:255–260

Zilles K, Amunts K (2010) Centenary of Brodmann's map – conception and fate. Nat Rev Neurosci 11:139–145

Zweig RM, Ross CA, Hedreen JC (1988) The neuropathology of aminergic nuclei in Alzheimer's disease. Ann Neurol 24:233–242

MIX
Papier aus verantwortungsvollen Quellen
Paper from responsible sources
FSC® C105338

Printed by Books on Demand, Germany